Rethinking Biology

Public Understandings

Rethinking Biology

Public Understandings

Editors

Michael J Reiss
University College London, UK

Fraser Watts
University of Lincoln, UK

Harris Wiseman
Campion Hall, University of Oxford, UK

World Scientific

NEW JERSEY · LONDON · SINGAPORE · BEIJING · SHANGHAI · HONG KONG · TAIPEI · CHENNAI · TOKYO

Published by

World Scientific Publishing Co. Pte. Ltd.

5 Toh Tuck Link, Singapore 596224

USA office: 27 Warren Street, Suite 401-402, Hackensack, NJ 07601

UK office: 57 Shelton Street, Covent Garden, London WC2H 9HE

Library of Congress Cataloging-in-Publication Data

Names: Reiss, Michael J. (Michael Jonathan), 1958–   editor. | Watts, Fraser N., editor. |
    Wiseman, Harris, editor.
Title: Rethinking biology : public understandings / editors, Michael J. Reiss,
    University College London, UK, Fraser Watts, University of Lincoln, UK,
    Harris Wiseman, University of Oxford, UK.
Description: New Jersey : World Scientific, [2020] | Includes bibliographical references and index.
Identifiers: LCCN 2019030148 | ISBN 9789811207488 (hardcover) | ISBN 9789811208263 (pbk)
Subjects: LCSH: Biology--Popular works.
Classification: LCC QH309 .R48 2020 | DDC 570--dc23
LC record available at https://lccn.loc.gov/2019030148

British Library Cataloguing-in-Publication Data
A catalogue record for this book is available from the British Library.

For any available supplementary material, please visit
https://www.worldscientific.com/worldscibooks/10.1142/11478#t=suppl

To Brian Heap, John Polkinghorne and the memory of Mary Midgley

Author Biographies

William Beharrell is a medical doctor currently working for the West Suffolk NHS Foundation Trust. Prior to medicine, he was the deputy Managing Director of a Prince of Wales charity called Turquoise Mountain. His role included healthcare, community development, livelihoods and education in Afghanistan and the Middle East. He is a former Pathfinder fellow with the Royal College of Psychiatrists with research interests in neuropsychology, phenomenology, and pre-modern conceptions of the body. He has degrees from Cambridge, where he studied medicine, and Durham, where he studied Arabic with Persian, and has spent much of his career overseas in the charitable sector.

Celia Deane-Drummond is Director of the Laudato Si' Research Institute, Campion Hall, Oxford University. She was Director of the Centre for Theology, Science and Human Flourishing at Notre Dame from 2015–2019. She holds two honours degrees, in natural science and theology, and two doctorates, one in plant physiology and one in systematic theology. Her research interests are in the engagement of systematic and moral theology and the biological sciences, including specifically ecology, evolution, genetics, animal behaviour, psychology and evolutionary anthropology. She is joint editor of the international journal *Philosophy, Theology and the Sciences*. A selection of her more recent books include *Christ and Evolution* (2009); *Creaturely Theology*, ed. with David Clough (2009); *Religion and Ecology in the Public Sphere*, ed. with Heinrich Bedford-Strohm (2011); *Animals as Religious Subjects*, ed.

with Rebecca Artinian Kaiser and David Clough (2013); *The Wisdom of the Liminal: Human Nature, Evolution and Other Animals* (2014); *Re-Imaging the Divine Image* (2014); *Technofutures, Nature and the Sacred*, ed. with Sigurd Bergmann and Bronislaw Szerszynski (2015); *Ecology in Jürgen Moltmann's Theology*, 2nd edition (2016); *Religion in the Anthropocene*, ed. with Sigurd Bergmann and Markus Vogt (2017); *A Primer in Ecotheology: Theology for a Fragile Earth* (2017); *Theology and Ecology Across the Disciplines: On Care for Our Common Home*, ed. with Rebecca Artinian Kaiser (2018); *Evolution of Human Wisdom*, ed. with Agustin Fuentes (2018); *Evolution of Wisdom Volume 1: Theological Ethics Through a Multispecies Lens* (2019); and *Virtues and the Practice of Science*, ed. with Thomas Stapleford and Darcia Narvaez (2019), https://virtueandthepracticeofscience. pressbooks.com/.

David J. Depew is Emeritus Professor of Communication Studies and POROI (Project on the Rhetoric of Inquiry) at the University of Iowa. He was previously Professor of Philosophy at California State University, Fullerton. Much of his work is in the philosophy, history, and rhetoric of evolutionary biology, writing often with Bruce H. Weber. He also writes on ancient biology and its relation to modern; and on how rhetoric, philosophy, and science have interacted since antiquity. His most recent book is *Darwinism, Democracy, and Race*, with John P. Jackson (Routledge, 2017).

Ilya Gadjev is an honorary Senior Research Associate at the UCL Institute of Education, University College London. Having completed a Marie Curie post-doctoral fellowship in genetics and sustainability he remained in Cambridge where, after a period at the United Nations Environment Programme — World Conservation Monitoring Centre and various engagements with the University and other organisations, he developed a scholarly interest in the relationship between science and society. Over the last years, Ilya has led and contributed to a number of projects in the spheres of life and environmental sciences, sustainable development, science outreach, and education. His publications cover topics ranging from molecular biology and physiology to philosophical problems in biology and the public understanding of genetics.

Derek Gatherer was Lecturer in Molecular Genetics at Liverpool John Moores University from 1996–1999, and has been a Lecturer in the Division of Biomedical and Life Sciences at Lancaster University since 2013. During the intervening period, he worked in the pharmaceutical industry and for the Medical Research Council Virology Unit. Beginning as a geneticist in Glasgow in the early 1980s, then a molecular embryologist during subsequent stints in London, Quito, Warwick and Cambridge, he moved rapidly through the evo-devo movement into a more general and philosophically inclined evolutionary theory during his time in Liverpool, then onto computational biology while in the pharmaceutical industry and the MRC. The apparent conviction of many systems biologists that they are now no longer reductionists but holists has spurred him into a new research interest about exactly what this means.

Francis Gilbert is Professor of Ecology, Faculty of Medicine and Health Sciences at the University of Nottingham. He studies the evolution of ecological and behavioural attributes of organisms, mainly in the field. He is particularly interested in the evolution of life histories and mimicry in insects, coevolution between plants and insects (especially in pollination), and in the importance of habitat fragmentation to populations and conservation.

Niels Henrik Gregersen is Professor of Systematic Theology at the University of Copenhagen. His primary research interests include Science and Religion, especially philosophical issues and theological issues regarding the significance of evolutionary theory and complexity studies; and Contemporary Theology, especially concepts of God in 20th century theology, deep incarnation, and eschatology. Also philosophical and social anthropology, concerning risk and risk-taking, gift and generosity and the reception of Luther and Reformation theology in the 19th and 20th century.

Richard M. Gunton lectures in statistics at the University of Winchester and recently completed a fellowship with the Centre for the Evaluation of Complexity Across the Nexus. He has undertaken fieldwork in South Africa, Australia and the UK and has published widely in the fields of ecology and nature conservation, with an increasing focus on philosophical issues. He is an Honorary Senior Research Associate at University College London, a member of the British Ecological Society and a Fellow of the Royal Geographical Society. He also coordinates the Faith-in-Scholarship initiative of the Leeds-based Thinking Faith Network.

Elizabeth Jones is a Historian of Science with a PhD in Science and Technology Studies from University College London. She specialises in the contemporary history of palaeontology as well as ancient genetics and genomics. Her research also overlaps with work in science communication studies, media studies, and celebrity studies. Currently, Jones is a Research Adjunct in the History of Science Lab at the North Carolina Museum of Natural Sciences in Raleigh, and she is writing her first book, *Ancient DNA: The History and Celebrity of a Science in the Spotlight.*

Ottoline Leyser received her BA (1986) and PhD (1990) in Genetics from the University of Cambridge. After post-doctoral research at Indiana University and Cambridge, she took up a lectureship at the University of York, where she worked from 1994–2010. Among her honours are the Society of Experimental Biology's President's Medal (2000); the Royal Society Rosalind Franklin Award (2007); the International Plant Growth Substance Association's Silver Medal (2010); and the UK Genetics Society Medal (2016). She was appointed a Dame Commander of the Order of the British Empire (DBE) in the 2017 New Year Honours list. She is a Fellow of the Royal Society, a Foreign Associate of the US National Academy of Sciences and a Member of the European Molecular Biology Organisation and the Leopoldina. She is Chair of the British Society for Developmental Biology, and of the Royal Society's Science Policy Advisory Group. She is Co-Editor in Chief of *Current Opinions in Plant Biology* and an Editor of *Development*.

Michael J. Reiss is Professor of Science Education at UCL Institute of Education, University College London and a Fellow of the Academy of Social Sciences. He has written extensively about curricula, pedagogy and assessment in science education and has directed a very large number of research, evaluation and consultancy projects over the past twenty years funded by UK Research Councils, Government Departments, charities and international agencies. He was a member of the Farm Animal Welfare Council/Committee (2004–2012); Director of Education at the Royal Society (2006–2008); a member of the GM Science Review Panel (2002–2004); a Specialist Advisor to the House of Lords Select Committee on Animals in Scientific Procedures (2001–2002); and is a member of the Nuffield Council on Bioethics (2019–present).

Michael Ruse is a philosopher of science who specialises in the philosophy of biology and is well known for his work on the relationship between science and religion, the creation — evolution controversy, and the demarcation problem within science. He currently teaches at Florida State University. He was born in England and took his undergraduate degree at the University of Bristol (1962), his Master's degree at McMaster University, Hamilton, Ontario (1964) and his PhD at the University of Bristol (1970). He is a Fellow of the Royal Society of Canada.

Mark Vernon is a writer, broadcaster, psychotherapist and former Anglican priest. He contributes to programmes on the radio, writes and reviews for newspapers and magazines, gives talks and podcasts. His previous books have covered themes including friendship and God, ancient Greek philosophy and wellbeing. He has a PhD in ancient Greek philosophy, other degrees in physics and in theology, and works as a psychotherapist in private practice. For more see www.markvernon.com

Fraser Watts is Visiting Professor in Psychology of Religion at the University of Lincoln, and a former Reader in Theology and Science at the University of Cambridge, where he was Director of the Psychology and Religion Research Group and a Fellow of Queens' College. He is a former President of the British Psychological Society and of the International Society for Science and Religion, and former Chair of the British Association of Christians in Psychology. He remains Director of the Cambridge Institute for Applied Psychology and Religion, and is Executive secretary of the International Society for Science and Religion. His recent books include: *Living Deeply: A Psychological and Spiritual Journey* (Lutterworth, 2018); *Psychology, Religion, and Spirituality* (CUP, 2017); *Evolution, Religion and Cognitive Science: Critical and Constructive Essays* (edited with L Turner, OUP, 2014); *Head and Heart: Perspectives from Religion and Psychology* (edited with G Dumbreck, Templeton Press, 2013); *God and the Scientist: Exploring the Work of John Polkinghorne* (edited with C Knight, Ashgate, 2012); and *Spiritual Healing: Scientific and Religious Perspectives* (edited, CUP, 2011).

Harris Wiseman is a Research Fellow at Campion Hall, the University of Oxford. He has a PhD in Divinity (psychology of religion) from the University of Cambridge, where he was Research Associate for two years. During that time he published *The Myth of the Moral Brain — The Limits of Moral Enhancement* (MIT Press). His bioethics work has been published in many of the leading journals in the field, the *American Journal of Bioethics Neuroscience, Cambridge Healthcare Quarterly*, the *Royal Institute of Philosophy, Bioethics, and Zygon*. He is convener of the Boyle Lecture Series; honorary Senior Research Associate at the Institute of Education, UCL; and remains a regular contributor to the Geneva Center for Security Policy's *Geopolitics and Global Futures* programme, lecturing on the neurophilosophy of global security.

Steve Yearley is Director of the Genomics Forum (officially the ESRC Genomics Policy and Research Forum) at the University of Edinburgh. He is Professor of the Sociology of Scientific Knowledge at the University of Edinburgh and — since 2017 — Director of the Institute for Advanced Studies on the Humanities (IASH). He is primarily interested in social studies of science and in environmental sociology and is particularly concerned with areas where these specialisms overlap: for example, in environmental controversies with a pronounced scientific element (such as with recent disputes over the safety or otherwise of GMOs and the emerging concerns around synthetic biology) or, for example, in attempts to foster public engagement in technical decision making in environmental areas (for instance, through his work on citizen engagement in urban air quality issues). He has been closely involved — initially through the Wellcome Trust — with work on social aspects of human genetics and with social science questions relating to bioethics.

Preface

This book arose from a concern about how biology is understood in the public domain, and a sense that there is a growing discrepancy between how biology is actually developing and how it is understood by the public. The trend in biology is towards increasing recognition of complexity. Almost everywhere you look, it is clear that plants and animals are shaped by many different causal factors that are interacting in complex ways. But the public is often given a very simplified version of this complexity, one that emphasises single causal factors.

The authors of this book suggest that this is due partly to biologists often oversimplifying matters when they present them to the public, and partly due to the requirements of eye-catching media. But the result is significant *mis*communication of biology, which in turn engenders various assumptions that have little foundation in research. The hope is that this book goes some way to correcting these oversimplifications by presenting some of the exciting complexity of biology in accessible language. The authors also reflect on some of the reasons why misunderstandings about biology are gaining ground.

All the chapters here arise from a research project funded by the Templeton World Charity Foundation, to which the authors express our gratitude. Many authors were core members of that project on 'The New Biology: Implications for Philosophy, Theology & Education', and all were associated with it in some way or other. The result is that all the contributors have had the opportunity to discuss things together much more intensively than is typically the case with edited books. That is not to say that there was unanimous agreement about everything, though there is an impressive meeting of minds. However, all of the authors have had the opportunity to refine our ideas in the context of a series of passionate and exciting exchanges that were hugely stimulating, and hope some of that shines through in the chapters that follow.

Michael J. Reiss

Fraser Watts

Harris Wiseman

December 2018

Contents

Part One

Introducing the Ideas

1 Introduction: Rethinking Biology

Fraser Watts*

Biologists always need to grapple with integrating two explanatory approaches. On the one hand there is necessarily an effort to drill down to the lowest possible level to explain what is happening in whatever is being studied. That involves looking at how higher level processes arise from lower level ones. On the other hand, there is a need to consider how the broader context influences bottom-up processes; that involves looking at how the whole influences the parts. Neither approach is satisfactory on its own. There is always a need to integrate the consideration of how parts influence wholes with how wholes influence parts.[1]

This book arises from a concern that in the public dissemination of biology the need to integrate these different perspectives is not coming across well. In popularisations, simplistic micro-explanations always seem to arouse most interest and to capture the headlines. That risks distorting and simplifying the complexity of biological processes, and can mislead people. In this book we are urging a concerted attempt to come to grips with the interactive complexity of biology, and to find ways of conveying it to the public accessibly and effectively.

*Visiting Professor in Psychology of Religion, University of Lincoln, Brayford Pool, Lincoln LN6 7TS, UK.

We are particularly concerned with how biology is communicated to the public. Too often, what comes over to the public is a crude, out-of-date, simplistic, monocausal, reductionist biology. Why so? Why is biology so misrepresented? Who is responsible? It is partly the media, of course, but I suggest that biologists themselves are often partly responsible. When it comes to communication with the public, they tend to oversimplify in a way that distorts.

Simple discoveries are obviously easier to popularise than complex ones. But I think there is more going on. It is also that many people find it exciting to believe that scientists are getting to grips with nature, stripping away the veils and making nature reveal her secrets. Mary Midgley has pointed out how remarkably sexist this kind of rhetoric is;[2] it has been prominent from the 17th century onwards, and indeed throughout the modern period. The idea that we have almost got nature sorted out is one that many people find exciting. So, to get their excitement, they constantly exaggerate how far we have got with such scientific exploration.

There may sometimes be a religious agenda operating here, or rather an anti-religious agenda. Sometimes scientific ideas become popular because they point in an atheist direction. I think the current popularity of the idea of 'many worlds' (the multiverse) is partly because it seems as though it blocks a design argument for God breaking out again. I suspect that the appeal of simplistic, deterministic biology for some people arises from their hunch that it points in a non-religious direction. Lewis Wolpert, a brilliant populariser of biology, illustrates the alliance between reductionist biology and atheism,[3] an alliance that I believe is unnecessary.

Arthur Peacocke, a biochemist turned theologian, was one of the first to recognise that biology was shifting in a more holistic direction, and to see the implications of that for a religious world view. He began by talking about levels of explanation, emphasising the importance of top-down as well as bottom-up explanations. Later, he shifted to the vocabulary of wholes and parts, talking about 'whole–part constraint'.[4] The effect of top-down or whole–part influences in biology can provide an analogy for how the everyday world could be influenced by a broader spiritual context (or 'God'), that is the ultimate 'whole' of which our everyday world is a part.

Parts and Wholes

It was one of the great achievements of biology as it emerged as a scientific discipline in the late nineteenth and early twentieth century to develop the concepts of cells and genes. We now often talk about the living world as though it was made up of specific bits, like genes, neurones or cells. But these are really all abstractions, or constructions. They are not building blocks, like little bits of hard matter. Even with cells, there is no really hard boundary, nothing as sharp as a cell wall. In reality, one thing flows into another; walls are permeable. We are dealing more with gradients than with walls.

If the way we talk suggests otherwise, it is just a language of convenience, and to some extent misleading. Our world, as Coleridge would have said, is one of distinctions rather than divisions. I agree with Coleridge that a lot of our philosophical problems arise from taking mere distinctions and treating them as divisions. The living world, at least, is not like that, and we confuse ourselves if we imagine it is. Cells and genes are very useful concepts in biology, but we mislead ourselves if we over-reify them.

The picture of the living world that is now emerging from biology is very interesting, and has broad implications. It is a picture of a world that is 'interdependent',[5] a world of mutual influence, in which influences go in both directions, with wholes arising from parts, of course, but also wholes influencing parts. Sharma argues that in biology there are 'no referents independent of terms' (p. 99). She illustrates this in connection with signal transduction, arguing that the standard way of talking about this assumes an unnecessary over-reification of cells as animal-like agents. The living world is not an atomistic world with discrete building blocks of hard matter. It is a soft world, of soft matter and soft boundaries.

Carl Woese[6] offers the attractive suggestion that organisms are 'resilient patterns in a turbulent flow' (p. 176). He puts forward the metaphor of a child playing in a stream in which there is an eddy in the flowing water. The child disturbs the flow by poking a stick into it, but the eddy reforms. Organisms, he suggests, are not fixed and inflexible, but resilient patterns, similar to eddies; organisms are stable, not because they are invariant but because they are resilient.

Ecology is also giving us a picture of a world in which species support each other, small species support larger ones, and larger ones sometimes support smaller ones. There is a to-and-fro between each species and the ecosystem in which it operates, with influences in both directions. In an ecosystem there are delicate balances; if one thing is disturbed it has far-reaching consequences for many other species. Biology retains a tendency to analyse this from the perspective of one species at a time, and to have a less developed understanding of how the ecosystem as a whole operates.

People sometimes talk about 'holistic' biology. I want to be careful not to suggest that there is a separate biology called 'holistic' biology that proceeds alongside ordinary biology. There is just 'biology', and the task of all biology should be to integrate parts and wholes in whatever way is necessary to explain how biological processes operate to produce observable outcomes. I submit that all biology needs to consider wholes as well as parts, and the interaction between them, in as far as that is necessary to get a good model of what is going on. We are not advocating a special kind of biology called 'holistic biology'; we are advocating a concern with wholes in biology generally.

Advocates of a holistic approach to biology maintain that parts can only be understood in relation to wholes. Sometimes the relevant wholes are quite small scale, sometimes they are much larger scale. I am thus using 'whole' in a rather relativistic way. The relevant 'whole' varies from one situation to another. It can be just the immediate context of what is being studied, but still quite low-level. Sometimes it is the whole organism. For example, it seems that whether cells over-reproduce as they do in cancer depends in part on the overall immunological function of the body, not just on cells that have gone wrong.

Sometimes the relevant larger context is the environmental or ecological context in which the organism is living. How large a plant grows and where and in what direction it produces new stems depend on the context in which it is growing, not just on its DNA. There is nothing ideological or doctrinaire about what is meant by whole. The 'whole' that needs to be considered is simply the one that has most influence on whatever is being studied. Often there will be a series of wholes nested within one another.

Sometimes people talk rather about 'organismal' rather than 'holistic' biology. The organicist tradition is older and broader, and is more about root metaphors and conceptual frameworks.[7,8] In its radical form, organicists would say that everything in nature should be understood as a living system. 'Holistic' and 'organismal' approaches draw on different intellectual traditions, but I suggest that biologists who see themselves as holistic are probably also organismic, and vice versa. We are dealing more with different ways of conceptualising this approach to biology, and with different intellectual traditions, rather than with different groups of biologists.

Determinism, Reductionism and Mechanism

I now turn to issues about determinism, reductionism and mechanism in biology. It might be supposed that a holistic approach to biology makes it less deterministic, though that depends on what you mean by 'determinism'. Science always wants to provide models with good predictive value, and rightly so; that is in the nature of scientific investigation. However, there is a pernicious kind of determinism that wants to limit what variables are admissible in making predictions. At the extreme it becomes monocausal determinism, which tries to explain what will happen to one variable entirely on the basis of what will happen to another.

Common examples are the attempts to explain what people do in terms of genes or neurones. The monocausal determinism involved leads to slogans like 'my genes made me do it', or 'my neurones made me do it'. I am very critical of that kind of biological determinism. It turns genes and neurones into agents and anthropomorphises them. There are very few examples of monocausal determinism in nature, so the attempt to force nature into the monocausal mould is just bad science.

There sometimes seems to be a rampant monocausal determinism that does not fit the biological facts, but which is prominent in popularisations of biology.

Apart from a few exceptions, the search for monocausal reductionism is a hopeless project, because it doesn't reflect how the living world works. An adequate kind of determinism in biology will need to take a rich range of variables into account, including both micro variables and more holistic and contextual variables, and it will need to consider how they interact. If that leads to exact prediction, well and good; then we will have reached a convincing kind of determinism.

Ultimately, the question of whether the biological world is entirely determined is metaphysical and beyond the scope of biological inquiry. What matters is that biologists produce models capable of describing what they are studying, that explain what is going on, and which make accurate predictions. The notion of biological certainty is a myth that does not accurately describe what biologists set themselves to do.

There are similar issues about reductionism. Again, there are different kinds of reductionism. Dan Dennett[9] has distinguished between good and greedy reductionism. Carl Woese[6] makes a similar distinction between empirical (methodological) reduction and fundamentalist reductionism. The search for exact prediction is sometimes motivated by a reductionist agenda. Exact prediction seems to allow the reductionist move that says that: because one thing can be explained entirely in terms of another, the thing that is explained is not what it appears; it is just some kind of epiphenomenon, one thing masquerading as another. For example, in this view people can be described as being 'just' (or 'nothing but') a bunch of cells, which are all similarly just a bunch of interacting elementary particles.

It is worth emphasising here that complete reductionism depends on getting *complete* explanation, reflected in fully accurate prediction. I want to emphasise how rarely that condition is met in biology. That ought to mean that reductionist moves are seldom made, but the problem is that reductionists often don't wait until they have got a complete explanation. They tend to make reductionist moves ahead of time, on the strength of a belief that they will eventually get a complete explanation. This is a kind of cheating, and I think we perhaps need to be more vigilant for it, and to blow the whistle on it sooner.

If you meet anyone wanting to make reductionist moves in biology, my advice is to challenge them to produce their complete explanation. If they can't, they should be advised to hold off with their reduction until they have actually got a complete explanation. Reductionism is usually attempted on a wing and a prayer, and without the necessary conditions being met. It should be one of the benefits of a move towards a more holistic approach to biology that there will be growing recognition that complete explanation in biology is usually 'pie in the sky'.

The question also arises of what place there is for the search for biological 'mechanisms' in a more holistic biology, and again it depends on what you mean by 'mechanism'. There are two rather different meanings of 'mechanism' in science, as has been pointed out by Rom Harré.[10] It can be either a mechanical contrivance, *or* a detailed explanatory model, i.e. 'any kind of connection through which causes are connected' (p. 118), as Harré puts it. There is nothing in a more holistic approach to biology to deter or discourage the search for detailed explanatory models. But it does warn that these models, if they are to be at all adequate, are unlikely to take the form of mechanical contrivances.

Carl Woese[6] sets out some of the key differences between machines and organisms. Unlike machines, organisms have parts that continually renew; an organism's stability arises from its resilience; it is more than a collection of parts and has 'a sense of the whole' (p. 176). Mechanisms in the broad sense are good and helpful but, in biology, the mechanisms are not likely to be really mechanistic. Good models often have a degree of verisimilitude, and models in biology that take the form of mechanical contrivances are going to lack that.

Shifting Emphases in Biology

Over the decades, biology has oscillated over the relative importance it has attached to wholes and parts. At the risk of oversimplification, biology was probably more concerned with wholes in the first half of the twentieth century than in the second half. The years after WWII were dominated by striking advances in molecular biology, which made biology as a whole more enamoured of the power and scope of micro explanations. There are times when it helps to make scientific progress to focus down and ignore other factors for the time being; but there comes a time when you have to broaden out and take those broader factors into account again. Molecular biology began with great promise and has undoubtedly accomplished a great deal. However, as Carl Woese[6] has argued, it has hard now to resist the conclusion that, in its original form, it has run its course, and that some broadening of the paradigm is now needed.

One of the great achievements of molecular biology was undoubtedly the discovery of the structure of DNA. However, as Woese argues, there was a tendency to exaggerate the significance of that discovery. It is one thing to discover the structure of DNA, but quite another to understand how that translates into an organism. We are now very aware that organisms are not wholly determined by their DNA but depend on interactions with a wide range of more contextual, holistic factors. In recent decades there was been a swing back towards taking such factors into account more fully. Those influenced by Hegel might see this as an example of the classic sequence of thesis, antithesis and synthesis; with a holistic approach as the thesis, and the period dominated by biochemistry as the antithesis. What is needed now is not just a swing back of the pendulum, but a proper integration of micro explanations and more contextual explanations.

There is sometimes talk about holistic biology being 'new' but, of course, holistic biology is not really new. Holistic approaches to biology go back at least to the early 20th century, and the organicist tradition is even older. However, I think it is correct to say that something has been changing in biology in the last decade or two. Michael Ruse has provided a valuable history of holistic biology,[11] illustrating the continuity between current and older holistic trends.

One of the important driving forces in the move away from micro explanations has been epigenetics. The core idea of epigenetics, that non-genetic factors affect how an organism's genes express themselves, is not new. It is a term coined by C. H. Waddington during WWII,[12] though he may have meant something slightly different by it from what is meant now. But there is new evidence to support epigenetics, and a detailed understanding of how it works, which has brought it back into greater prominence. There have been other significant developments in biology that have also nudged biology in a more holistic direction, including the extended synthesis in evolutionary biology, and developments in ecology such as a better understanding of niche construction. In part this is a reversion to an older strand of biology.

There are particular puzzles about what is, or should be, meant by 'systems biology', what it is, and whether it is yet another term for the developments in biology in which we are interested here. It is a term that has been particularly popularised in the UK by Denis Noble.[13] In fact, the term 'systems biology' is used for very different things. For some people, it just means taking a lot of variables into account. That in turn, pushes you into mathematical modelling, as the only viable way of considering a large number of variables. However, there is often nothing particularly holistic, interactive or contextual in how those variables are considered. There is nothing wrong with considering a large number of variables, except that it may distract you from getting to grips with what is actually going on with any of them.

However, there is another kind of systems biology that plays close attention to how variables interact. For obvious practical reasons such work is often done on quite small-scale systems. There seems to be no generally agreed term for that kind of systems biology, though 'integrative biology', with its emphasis on the integration of the contribution of different disciplines, comes close. We are very interested here in that kind of systems biology, but it is very different from much of what is called systems biology.

The Structure of the Book

This book is organised into three sections. It begins with a short general section, of which this chapter is part. Next comes a chapter from Derek Gatherer, a systems biologist who is unusually interested in the history and philosophy of systems biology, and who examines the currently fashionable systems biology from that point of view. He identifies some of the more recent attempts to set out reductionism and holism in biology, taking a particular interest in the 'relational' biology of Robert Rosen.

Next comes a pivotal chapter by Ottoline Leyser and Harris Wiseman; Leyser is a distinguished British plant biologist, currently Director of the Sainsbury Laboratory in Cambridge. While avoiding a switch into a separate kind of holistic biology she sets out the compelling scientific case for why biology needs to take a range of contextual factors into account in explaining how organisms develop and function. Her focus is on plants but her general points apply to all biological systems.

Next comes a group of seven chapters that deal with specific topics in biology, many of them applying the general principles set out by Leyser. The first three are concerned with different aspects of evolution. Ilya Gadjev traces the shift from genetics to epigenetics, taking a particular interest in the broad social significance of that shift. David J. Depew looks at the way the modern synthesis in evolutionary theory is now coming under pressure, but argues for a revised and extended synthesis that takes a broader range of factors into account, rather than for an overturning of the modern synthesis. Elizabeth Jones and Michael Ruse then look at the implications for human evolution, and explore how DNA has captured the popular imagination. DNA is much more complex than is often realised, and recent research may challenge our concept of what it is to be human.

Harris Wiseman then examines current attempts to explain (or over-explain) human behaviour in terms of neuroscience, and how such explanations will always be more partial than is often realised; it is an issue that has broad social and legal ramifications. Will Beharrell then looks at medicine, and reflects on the fact that there is widespread acceptance of a broad bio-psycho-social-spiritual approach, but without much recognition of its implications. Medicine could be more effective if it went further in taking the whole-person context into account.

Richard Gunton and Francis Gilbert argue for the fundamental importance of ecology for all biology, and especially for evolution. Their claim is that ecology is foundational, in that evolution can't be understood without it. Finally, in this section, Michael J. Reiss looks at how diet has evolved over the centuries, and explores the concept of a 'good diet'. That turns out to be based on more insecure scientific foundations than might be supposed, and to be associated with questionable popular assumptions.

In the final section of the book, we look at various aspects of the social context of how biology is understood. Steven Yearly argues that biological and societal issues are closely connected, and that how the public understands biology has wide social significance; he also reviews recent evidence on the public understanding of biology. Michael J. Reiss emphasises the extent to which children have ideas about biology before they begin formal education on the subject, and that this complexifies the teaching of biology. He focuses particularly on how that is exemplified in children's understanding of evolution.

Niels Gregersen examines the implications of understandings of biology for how religion is understood. The systemic, contextualised way in which biology is now developing opens up more ways in which God's engagement of the world can be conceptualised. Next, Harris Wiseman looks at how science is understood in society, emphasising the problems that arise from regarding science as a highly technical activity carried out by experts, and arguing for the advantages of a fuller public engagement with science. Mark Vernon contributes a journalist's perspective on how biology is communicated to the public, reflecting on the pressures towards over-simplification and how these might be navigated. Finally, Celia Deane-Drummond assesses the various chapters of the book and presents an overview of their implications.

Edited books come together in a variety of ways, but this one arises out of close interaction over the three years of a research project on 'The New Biology: Implications for Philosophy, Theology & Education' (funded by the Templeton World Charity Foundation). Most of the contributors were fully involved in the project, and all were connected with it in some way or other. We hope this gives a coherence to our examination here of how biology is understood by the public, and the particular issues associated with simplistic, reductionist misrepresentations of biological research.

References

1. Watts F and Reiss MJ. (2017) Holistic biology; what it is and why it matters. *Zygon* **52**(2): 419–441.
2. Midgley M. (1994) *Science as Salvation: A Modern Myth and its Meaning.* Routledge, London.
3. Wolpert L. (2006) *Six Impossible Things Before Breakfast: The Evolutionary Origins of Belief.* Faber & Faber, London.
4. Peacocke A. (1999) The sound of sheer silence: How does god communicate with humanity? In: Russell RJ, Murphy N, Meyering TC and Arbib MA (eds.) *Neuroscience and the Person: Scientific Perspectives on Divine Action,* pp. 215–247. Vatican Observatory, Vatican City State.
5. Sharma K. (2015) *Interdependence: Biology and Beyond.* Fordham University Press, Fordham.
6. Woese CR. (2004) A new biology for a new century. *Microbiol Mol Biol Rev* **68**(2): 173–186.
7. Ruse M. (2010) *Science and Spirituality: Making Room for Faith in an Age of Science.* Cambridge University Press, Cambridge.
8. Ruse M. (2013) *The Gaia Hypothesis: Science on a Pagan Planet.* Chicago University Press, Chicago.
9. Dennett D. (1995) *Darwin's Dangerous Idea: Evolution and the Meanings of Life.* Simon and Schuster, New York.

10. Harré R. (1972) *The Philosophies of Science*. Oxford University Press, Oxford.
11. Ruse M. (2017) The Christian's dilemma: Organicism or mechanism. *Zygon*, **52**(2): 442–467.
12. Waddington CH. (1942) The epigenotype. *Endeavour* **1**(i): 18–20.
13. Noble D. (2006) *The Music of Life: Biology Beyond the Genome*. Oxford University Press, Oxford.

2 Integrative Biology: Parts, Wholes, Levels and Systems

Ottoline Leyser* and Harris Wiseman[†]

I would like to preface this chapter by noting how interdisciplinary dialogues make me feel rather inadequate.[a] Certainly, I have no expertise in the vast majority of the disciplines being talked about in this book. I have no deep knowledge about the history of biology, nor about the philosophy of biology. How I can serve here, if you will, is as a data point. I am someone who does science, and I hope what I say here can be used as data for these interdisciplinary analyses. So, what kind of science do I do? I am interested in plants. I am a developmental geneticist, and I am therefore interested in understanding how a single cell, in this case a plant cell, manages to become a complex multicellular organism, with all the cells in the right place, and at the right time, doing the right things to develop into and function as a complex, multicellular organism. It is an extraordinary question in biology, which I find completely gripping.

---

*Director & Professor of Plant Development, Sainsbury Laboratory, Cambridge University, Batemen Street, Cambridge CB2 1LR, UK.
[†]Research Fellow at Campion Hall, University of Oxford, Brewer St., Oxford OX1 1QS, UK.
[a]The 'I' throughout this chapter refers to the voice of Ottoline Leyser. The chapter was prepared with the help of Harris Wiseman.

The question of plant development is complex, and there are many layers to that complexity. But the important point to keep hold of is that this question is definitely *tractable*. It is problematic when people say of a biological problem: 'it's complicated', and use that as an excuse not to think about it. The complexity here is really interesting, and that's why thinking about it is so intriguing. And, as I say, the question is tractable. Biologists have made huge amounts of progress over the last 200 years in understanding this kind of question. How have we managed to do that? That's what I want to say something about now.

To explain all this, I am going to build up some kind of argument, a series of points, and I will do so by presenting things in a sequential way. I mention this because that process is itself a key tool for getting into complexity. What you need to do is to pick some bit of the puzzle, to start from there, and then to move forward in a sequential way. Building sequentially, you start to increase your understanding of that complexity. But, what needs to be kept in mind is that the complexity in the real process has nothing to do with that linear arrangement of ideas. So, that's the first key point, which we'll come back to, and it will be a recurring theme for this chapter. Namely, that there are many things we do in order to get our heads round the complexity of biology that are, at once, key tools for comprehending complexity, but are fundamentally contrary to the kind of complexity we are trying to study. *And that is not a problem.* Indeed, that's at the core of the message I want to communicate.

In talking about biology, you hear an awful lot of 'isms' getting discussed. My favourite 'ism' is *pragmatism*. You do what you need to do in order to get the job done. And one of the things you need to do in order to study complexity is to linearise a non-linear system. Biology, essentially, is about how living systems work. I am going to use an example which comes directly from my own research: Arabidopsis plants, as that's the model system I work on. Arabidopsis is the equivalent of the fruit fly for plant biology.

Arabidopsis is a small weed you have probably seen growing in the cracks between paving stones. As it grows, it first produces a rosette of leaves, and then it sends up an elongating shoot. Flowers can be found at the top, from which form the fruits containing the seeds. And, after the plant flowers, it branches. Arabidopsis makes branches in a top-down sequence. Branches grow from buds at the base of the leaves. The way this whole process happens is that, at the tip of each growing shoot, there is a group of cells called the meristem. In animal biology they would be called 'stem cells', but because plants have stems this term can be rather confusing. So, we often call this group of cells the meristem. The meristem builds the plant underneath it — it makes leaves at its flanks, and at the base of each leaf there's another meristem. Those meristems typically become dormant as a bud, after making a few leaves, but these buds can reactivate to make a branch. The main meristem can inhibit the activity of the secondary meristems. But, if you chop off the top of the plant and take that meristem away, the secondary meristems will activate and make branches. As gardeners will know, this is why pruning roses makes them bushy. So, the meristems essentially talk to each other, and they make collective decisions about who should activate, and who should not activate. And how that happens is the kind of question I am interested in addressing.

Damage to the main shoot tip by a pair of secateurs, or by a herbivore, are some of the things that can come in from the environment to influence the decision about whether or not meristems activate. The other factor that we work on is nutrient availability, for example through fertilisers. If you take two genetically identical plants, and feed one with full nitrogen fertiliser, while the other is grown with limited fertiliser, you find that the latter makes far fewer branches. Some people think this a boring and obvious result. They are assuming that the fewer branches are simply a result of there being less nitrogen to make more branches, however this is not true. What is happening is that in underfed plants, buds in the base of the leaves are suppressed, and any nutrients that are available are being invested in building roots.[1] If a plant is limited with respect to nutrient availability then diverting any nutrients that are available to producing roots is a good strategy. That way, the plant can find more nutrients. These are really quite sophisticated decision-making processes that plants are using to determine where to invest their current resources to optimise their chances of future resource maximisation.

Some of the decisions here involve epigenetic changes. Epigenetic changes are often central in driving developmental programs forward more generally, and they are based on the genotype of the plant, the environment in which it is growing, and the previous developmental decisions the plant has made.[2] So, there is a really rich collection of information sources that are being integrated together throughout a plant's life cycle, and biologists are particularly interested in understanding how that integration happens. In my case, I am interested in this integration with respect to the action of the meristems at the base of each leaf and whether or not they should grow out to make a branch. I use factors like nitrogen availability as an easy-to-manipulate variable, one that changes the meristems' decisions.[3] In manipulating things in this way, I hope to understand how plants then make those decisions.

Thinking about this more generally, in approaching any complex biological question, you come at it in two ways that are interconnected — *description* and *perturbation*. The information you get from these two interconnected sources are how you then pick the system apart. It should be noted that the very first thing you have to do is to describe what you're looking at. Description is absolutely essential. You have to give names to the phenomena, the 'bits' and 'things' that the plants are doing, because otherwise you cannot talk about it (indeed, it can be difficult even to think about it without this process of naming of parts and processes). *But those parts and processes that you name are inherently contingent on what you happen to be thinking about at the time.* This descriptive process does not end there. But, once a thing has been named, then that is end of it, is it not? No — because your description of the biological process in question is heavily biased by what you think might be going on. The parts you pick to name, and the names you give them, are contingent on how you are thinking about the question at that moment. And that is subject to change.

If there is a new biology at all (and I'm a bit sceptical about that, because I think biology is about understanding how living systems work — it always has been, and always will be), then the newness here comes from the fact that we've had a really rapid acceleration in the tools available to us for describing systems. That acceleration in the development of tools has allowed us to describe parts that we have not been able to capture before. That has been revelatory, and the ability to describe, capture, and see what's going on, is profoundly tied in to the questions we can ask. I will come back to that at the end of the chapter.

What I want to emphasise here is that the sort of complexity we have been trying to think about so far in this chapter refers to a system that is operating at a very wide range of *scales*. This complexity over a range of scales is a really interesting and exciting aspect of biology — particularly if you are a developmental biologist and geneticist. To repeat, even though I have to use the word 'complexity' a great deal, I want to emphasise that complexity does not give people an excuse to avoid thinking about things. In fact, the opposite is true.

To illustrate the multi-scale nature of this complexity, let us consider an Arabidopsis plant at a series of scales, and pay attention to how each scale influences how we give names to the 'bits' and 'parts' of the plant.

Arabidopsis plants are quite small. Their rosettes are a few centimetres in diameter, and their flowering shoots are maybe 30 cm high. The convention is to call this thing, along with its less visible root system, a 'plant'. We have decided that it is one single distinct plant, despite the fact that it is living in a broad environment, and interacting with all kinds of microbes in the soil, and so on. Zooming in, we have also named some parts of the plant. For example, there are leaves, which in Arabidopsis are maybe 1 cm long. Zooming in a little bit more, we can see on the surface of the leaf some interesting jigsaw puzzle-shaped things that we've decided to call 'cells', which are maybe 100 μm across. On the leaves there are also beautiful projecting hairs we have decided to call trichomes.

And, you can keep zooming in. Looking more closely at the cells, and we've decided that the cellulose-rich material bounding each cell should be called 'cell walls'. Inside the cell are some spheroids we have decided to call 'chloroplasts' where we have discovered that a process we have decided to call 'photosynthesis' happens. Another typically spherical structure inside cells we have called the 'nucleus', which is where most of the stuff we have called 'DNA' is. There are nuclear pores, connecting the nucleus to the surrounding cytoplasm. And inside is chromatin, which is a folded-up version of the DNA packed up by interactions with some specific proteins. And then, of course, there is the DNA itself, the impressive double helix that we all know and love. Therein are the bases that make up the genetic code, the As, Gs, Cs and Ts, pairing across the double helix. Now we're down to the nanometre scale.

With that, we've covered eight orders of magnitude. Just as a point of orientation, we can note that one meter with respect to the whole world is fewer orders of magnitude than the levels over which we have just zoomed in. What is really exciting here is the *relationship* between those tiny nanoscale events going on in and around your DNA for example, and the meter or centimetre parts at organismal level. We can describe biological systems at all those scales quite accurately now because of the tools we have available.

However, description is not enough to understand what is going on. We need to be able to poke these systems, as it were, and see what happens. So, the other major approach we use is *perturbation*. What I described above in terms of changing nutrient availability is a simple way of perturbing the Arabidopsis system. Essentially, we have got the same system with both plants, but I've given one of them lots of nitrogen, and given another one less nitrogen, and as a result some changes have happened. So, I can describe the differences between the plants, and that can give me some clues as to what's going on. Being a geneticist, my preferred kind of perturbation is, of course, genetic perturbation. The classical approach to genetic perturbation is essentially random. This is, in my opinion, a very powerful means of perturbation because it's less biased than the sorts of perturbation that you deliberately select to disrupt a given system.

Genetics works well as a way to unpick biological systems. You take an organism, in this case a whole tube of seed. You soak these seeds in a mutagen that changes DNA bases randomly. Thousands and thousands of seeds that you've treated in this way are then planted. And then, when they grow (or more accurately their offspring grow), the investigator looks out for ones that appear different in an interesting way. So, admittedly, there is some bias involved, because the investigator is making a choice about what he or she thinks is an 'interesting' mutation. That bias is dependent on the sort of question the investigator is asking. It is this question that guides the selection of the 'interesting' plants. For example, in my own research, I am asking how it is that plants manage to tune their branching according to the environment in which they are growing. The kinds of mutants I have picked look very branchy.[4–6]

This approach gives me two plants that are almost completely, but not wholly, genetically identical. They often differ genetically by one single base pair. For each of the interesting mutants we have chosen, we have spent a lot of time working out which base pair (or pairs) have been changed. To pick the example of one of our branchy mutants, although there is only one base pair difference, the mutant plant is much more highly branched than the plant line from which it was derived. With this mutant plant, there are lots of branches, and this is the case whether they are fully fertilised, or have only a limited nutrient supply. Essentially, we have broken the decision-making process through which the plants allocate their resources and therefore how branchy they become. In these mutants, they always branch whether or not they have enough fertiliser. We have done this by changing just one base pair in their DNA. The mutant plants, more or less, branch regardless of nitrogen availability. Almost every single one of their buds will activate, and that's that. The mutant plants can't respond to nitrogen supply anymore because of this one base pair change.

At this point, I could start talking about biological causality, but this is not very interesting for a variety of reasons. In particular, because the whole system is totally ridden with *feedback*, the very notion of causality is not so helpful. I am interested in how this system works precisely as a system, so I'm not really interested in exactly what is causing what. At the beginning of the lives of these two plants they were seeds, and at that point the only discernible difference was this one base pair (in every cell). Making an appeal to biological causality at this point is tempting, so you have to be careful. The temptation is to say: 'okay, we know what happens, the different base pair is what causes the plant to grow in this way', but that is not very informative. Instead, the interesting part is trying to understand *why* changing that one base pair has had this big impact on the way that the plants grow and develop.

We know quite a lot about why the sequence of bases in DNA is so important. This is encapsulated in the Crick central dogma, which states that DNA (in this case in the gene that we've changed) is responsible for making RNA, and that makes proteins, and then those proteins do something.[7] In this case, the particular gene that has been changed in the mutant plant encodes an enzyme, which performs an oxidation reaction that converts a compound called carlactone to carlactonoic acid.[8,9] The one base pair change means that this protein does not work anymore.

At this point, we have to take note of a key idea in genetics that is frequently missed in the discourse surrounding it. This is often framed in terms of a 'gene for' something or other, which is a seriously problematic framing for quite a lot of reasons. For example, people talk about the gene that 'causes' haemophilia. But there is no gene that causes haemophilia. Haemophilia arises from *not* having an active version of the gene that promotes blood clotting. So, it is not that one has 'the gene for' haemophilia, it's that one does *not* have the gene that would prevent you from having it. That is a key difference. The idea that there is a gene that gives you the disease is unhelpful. And that is what we have with the mutant Arabidopsis. A protein that is normally present and active is absent and therefore not active. The genetic approach allows you to break the system one gene at a time. So, if you take out the protein that makes the plant able to convert carlactone to carlactonoic acid, then you end up with a plant where almost all meristems activate to produce branches. And, one can do quite a good job at restoring the mutant plants back to their non-mutant form by resupplying the original version of the gene/protein involved or by giving them carlactonoic acid.

Even so, by focusing only on the protein level, we haven't yet got much beyond the nanoscale. We have not seen the *interaction* between the different scales. Just talking about proteins is not helping us understand how the decisions at the whole plant level are influenced by nanoscale events. To that end, a great deal of work has gone into trying to think about these processes as a *series of interlocking self-organising systems.* The current kinds of datasets that are being produced are now making reference to this kind of system-oriented terminology.[10] The most useful way of conceptualising what is going on, in my view, is to understand this Arabidopsis system as a series of dynamic, self-organising, feedback-driven systems, operating at different scales. Of course, in saying that, we find a lot that needs to be unpacked.

Let us look at some of the different levels and how they all, at each scale, can be described as dynamic self-organising systems of components at a smaller scale:

At the cellular level: there is a compound derived from carlactonoic acid acting as a signal, that is decoded by a signal transduction pathway. Through the action of this pathway, the carlactonoic acid derivatives, called strigolactones, are adjusting the levels of a small family of proteins. These proteins, among other things, are regulating the allocation of yet another family of proteins onto the membranes of cells in the plant, and these proteins are catalysing the export of a different compound, called auxin, out of cells. The amount of auxin inside the cells (which is now changing because of the presence of this transporter protein on the membrane) itself depends on several things. And the amount of auxin, working through different signal transduction pathways, is able to adjust the amount of carlactonoic acid and hence strigolactone. So, you can think of this cell-level network as a dynamic self-organising system of interacting molecules. This system is sensitive to both local and systemic inputs, and is therefore equilibrating, or shifting between equilibria, with different dynamics in different circumstances.

At the meristem tissue level: shoot meristems at the tip of growing shoots build the plant, producing stem beneath them and leaves at their flanks. You can think of this meristem as a dynamic self-organising system of cells, which is also sensitive to local and systemic inputs. Interactions between the cells set up patterns of expression of different genes across the meristem so that the cells specialise to fulfil specific functions. The cells in the middle of the meristem divide and maintain a population of mother cells that can keep the meristem going throughout the life of the plant. Cells at the flanks are competent to become leaves or stem, and more cell–cell communication, involving both chemical and mechanical signals between cells, establishes which cells get to be part of leaves and which cells become part of the stem. The process is dynamic. Cells produced in the centre of the meristem change their gene expression patterns as they get further and further away from the central region of dividing cells as a result of new cell divisions in the meristem creating more cells. At the meristem periphery, cells eventually get incorporated into leaves or stem, depending on interactions among the cells across the meristem tissues that ensure a balance between the two choices, in a pattern that spaces the leaves out around the stem.

At the plant level: each growing meristem exports auxin, which is transported down the plant and influences the activity of other meristems in the shoot. These interactions are also continuous and dynamic, so that if you prune away the leading shoot, removing it as an auxin source, buds further down the stem activate to produce branches, which in turn export auxin into the rest of the plant and prevent more buds from activating. So, you can think of the plant shoot as a dynamic self-organising system of meristems that are talking to each other.

This gives stacks of regulatory networks, or architectures, that are operating at different scales. Those scales all link together. They are not really separate or distinct, but thinking about them in that way is instructive for understanding how the system works. The natural human way of thinking about things seems to lead us toward seeing such processes as being both physically and dynamically linear, involving a sequential arrangement of parts or events. Indeed, it is very difficult to build an intuitive understanding of dynamic self-organising, feedback-driven systems, because humans think in a linear sort of way. Fortunately, through mathematically modelling these regulatory networks it is possible to layer up an intuitive understanding of the system. It also makes it possible to focus on different levels, and the interactions between them. In fact, all of our thinking about shoot branching in Arabidopsis is now absolutely dependent on computational and mathematical models. We have models that focus at the whole plant level, at the tissue level and at the cellular level. We pick the model most useful for addressing the question at hand. In the tissue level model, for example, we might have a term to represent how much strigolactone there is. It can be one simple equation linking strigolactone levels to auxin transporter levels. We do not include explicitly all the steps in between these two things, even though we know a lot about them and can model them with a smaller scale cellular level model if we want to. This entire cellular level model about the relationship between the two can be collapsed into a single summary equation that relates the two in the higher scale tissue level model.[11] In the end we should be able to bring all these models together through understanding how they feed into and out of each other.

Because of this, understanding how parts and wholes relate to one another is highly contingent. One person's whole is another person's parts. We see this with the process of transcription. Going back to the Crick central dogma, we described the process of DNA acting as a template to make RNA, which acts as a template to make proteins. The process of copying DNA into RNA is called transcription, and we know a huge amount about how it works. The enzymes that are required for transcribing DNA into RNA are collected at the right place on the DNA by proteins called transcription factors. They come together to mark the spot where these enzymes have to assemble. So, what we see is that this single word 'transcription' actually unpacks into a lower level dynamic self-organising network of interacting

DNA and proteins. I just call that network 'transcription', but some of my colleagues have devoted their lives to understanding the parts of this one whole that I simply call 'transcription'. *Their parts are my whole.* And this illustrates two points. First of all, each one of these scales can be thought of as a system, or as a part of a system, depending on which level you are looking at. The second point is that we have now given words not just to things and physical parts, but also to processes. Likewise, we can give names to entire sets of processes. So, the idea of wholes and parts breaks down in interesting ways.

In fact, the very manner in which you name your parts comes to influence the way you describe other parts and the way you understand the operations of the systems you are looking at. What's in a name? A great deal, it seems. We can see this influence with the concept of transcription. Crick's central dogma, for example, has consequences for how you think about the system. If you have hardwired this central dogma into your brain, that transcription is 'the main thing' to be concerned about, then you start interpreting your findings according to it, and you discover surprising things. You discover that there is an enzyme which does the opposite to transcription — copying RNA into DNA — so you call it 'reverse transcriptase', because as far as you are concerned, transcription is the primary process, and this reverse activity is running backwards and shouldn't be happening. You see this enzyme as a bizarre inversion of something fundamental, so you get very surprised. But the only reason you were surprised, and called it reverse transcriptase, is because you have already presupposed that transcription is the primary thing, and that is embedded in how you conceptualised the way that information flows in cells. So, the take home message is really that you have to define 'dogmas' like this, you cannot make progress unless you do so, but you have to keep in mind that you made them up as thought tools. You have to appreciate that your construction is bound to be wrong, and the words you've used, like 'transcription' (and therefore 'reverse transcription') are very much conditioned by the history of the subject. You have to keep your construction of the system fully open to challenge, otherwise it can constrain your thinking, which is a very bad idea.

Likewise, the whole idea of linearity in biological systems is problematic, because all biological systems that I have ever encountered involve various kinds of feedback within and between the various scales. For example, there is a classic 'negative feedback' process that operates across many biological systems. The amount of a substance that is made by a cell is regulated by how much of that substance is there already. The very thing being made limits its own production, which leads to an equilibration of the system at some particular point. Returning to the Arabidopsis plant, the amount of strigolactone in a cell depends on how much is being made, and how much is being degraded. The rate of production and the rate of degradation result in a dynamic equilibrium at a certain amount of strigolactone. The number of molecules of strigolactone in the cell stays the same, but they are different molecules from one moment to the next, as new ones are made and old ones are degraded. These sorts of systems are interesting because they can be easily retuned, for example in response to some change in the environment, like nutrient supply. This means that a lot of the interesting work now is happening at a level of thinking that is much more to do with dynamics and regulatory architectures than it is to do with 'parts' and what any of those individual parts do.

So, as well as this layering of scales, at each scale you can also layer in dynamics. There's 'the thing' itself, say, strigolactone; and then there's the flux of the thing. The flux of the thing is the rate at which it is being made and the rate at which it is being degraded. And that flux is tuneable by inputs and outputs that are dependent on the regulatory architecture of the system. So, it is complicated, yes, but it is tractable and it is beautiful.

In sum, we have to understand that our tools are just tools, and our dogmas and constructions are limited and bound to be wrong; but we shouldn't beat ourselves up about it either. They are necessary for making progress, so long as we keep in mind that they are bound to be limited and wrong, and keep up for grabs the names we give to physical parts and processes. Indeed, it is really important to remember that biological systems do not think they have parts. A plant has no idea what a cell is, no idea what DNA is — it's all just a 'bunch of stuff' hanging together in a state of dynamic flux. But we, as investigators, have to think that systems have parts otherwise we have no hope of understanding them, and no way of talking about them to one another (which is, of course, half the fun). And, again, one person's part is another person's whole; and one person's parts are another person's dynamic equilibria. Various investigators are operating at all these scales, with interests in different dynamic processes. At any one of these levels one can talk about parts, but then one can amalgamate those parts together and that thing becomes a whole at the next level. And similarly, every part, at whatever scale, is participating in a dynamic process, which can be understood within and across scales.

Moreover, because understanding systems involves talking about dynamics and interactions between the parts, we can go even further. Those interactions and the dynamics themselves can also be labelled as a kind a part. One of the interesting things happening in biology today is the labelling of regulatory architectures. For example, there are various different kinds of *switches* you can find in biological interactions. There are many different ways of switching from one process to another, for example, gradually, in a way that one can easily reverse, or very suddenly in a way that is hard to reverse. We are now beginning to invent words to describe those sorts of switching mechanisms. This moves biology into the realm of multi-order descriptions. Not only physical parts, but also these dynamic 'parts' can be stacked across scales. We can do all this because we now have new technologies, and these help us to see more deeply. If anything, that is what is new in biology: new technologies that are allowing us to see more deeply, and therefore moving us forward into this higher order of understanding. We can think about the properties of the equilibriated states and the ways to switch between them, rather than static physical parts.

If we are going to talk about reductionism in biology, pragmatism must be kept in mind. Thinking about parts is not really reductionism — the whole point is to use that as a tool for understanding how the system works. True reductionism involves picking a tractable question. You cannot possibly address the systems that we have been discussing here head on. You have to pick a bit with respect to which some progress can be made. The investigator hopes that he or she can pick a bit of what is going on that makes for a really good example, something that can be rolled out to help understand that larger whole. This choice is guided by the tools available at the time. That is the key, I think, to reductionism in biology — it is picking a tractable question in relation to the tools that are available, tools that will allow you to make progress in understanding things better. And if there is a 'new biology', it is that the questions that are now tractable are more interesting than they were a few years ago, which is of course, always the case, but maybe the rate of change now is higher. Throughout, we are trying to describe non-linear feedback mechanisms in linear ways. And (I hope the fact that I have laid out this chapter in a linear way will have made this point easier to understand), *it is valuable to linearise complex material*, as that is the only way to make progress in understanding complexity. To understand biology in any depth, you have to go with the way you can think most easily, and that is in terms of linear interactions between parts. And that is useful, but only if you know all the time that it's wrong. That's the key to keep in your head: *it will definitely not be right, but it will definitely be useful*.

References

1. De Jong M, George G, Ongaro V, *et al.* (2014) Auxin and strigolactone signaling are required for modulation of Arabidopsis shoot branching by N supply. *Plant Physiol* **166**: 384–395.

2. Lämke J and Bäurle I. (2017) Epigenetic and chromatin-based mechanisms in environmental stress adaptation and stress memory in plants. *Genome Biol* **18**: 124.

3. Leyser O. (2009) The control of shoot branching: An example of plant information processing. *Plant Cell Environ* **32**: 694–703.

4. Stirnberg P, van de Sande K and Leyser O. (2002) MAX1 and MAX2 control shoot lateral branching in Arabidopsis. *Development* **129**: 1131–1141.

5. Sorefan K, Booker J, Haurogné K, *et al.* (2003) MAX4 and RMS1 are orthologous dioxygenase-like genes that regulate shoot branching in Arabidopsis and Pea. *Gene Dev* **17**: 1469–1474.

6. Booker J, Auldridge M, Wills S, *et al.* (2004) MAX3/CCD7 is a carotenoid cleavage dioxygenase required for the synthesis of a novel plant signalling molecule. *Curr Biol* **14**: 1231–1238.

7. Crick FHC. (1958) On protein synthesis. *Symp Soc Exp Biol* **12**: 138–163.

8. Booker J, Sieberer T, Wright W, *et al.* (2005) MAX1 encodes a cytochrome P450 family member that acts downstream of MAX3/4 to produce a carotenoid-derived branch-inhibiting hormone. *Dev Cell* **8**: 443–449.

9. Abe S, Sado A, Tanaka K, *et al.* (2014) Carlactone is converted to carlactonoic acid by MAX1 in Arabidopsis and its methyl ester can directly interact with AtD14 in vitro. *Proc Natl Acad Sci USA* **111**: 18084–18089.

10. Milo R, Shen-Orr S, Itzkovitz S, *et al.* (2002) Network motifs: Simple building blocks of complex networks. *Science* **298**: 824–827.

11. Prusinkiewicz P, Crawford C, Smith R, *et al.* (2009) Control of bud activation by an auxin transport switch. *Proc Natl Acad Sci USA* **106**: 17431–17436.

3 Modelling versus Realisation: Rival Philosophies of Computational Theory in Systems Biology

Derek Gatherer*

From Biochemistry Through Molecular Biology to Systems Biology

One rainy spring morning in Glasgow in about 1984, I sat in a lecture theatre and heard my physiology professor tell the class that the problem with biochemists was that they just wanted to put everything in a bucket, blend it to a puree, and then talk about the properties of the resulting sludge. The class laughed, of course, not realising that this was physiology's oldest joke, possibly around 100 years old, and that our professor had been using it for almost as long. About eight years later, by which time I had become a lab research assistant at the University of Warwick, I heard just one of these 'bucket biochemists' complain, with somewhat less humorous intent, that it was virtually impossible by then to obtain a grant for doing biochemistry unless some gene-centred molecular biology angle could be found on the project. A quarter century later still, it is now the molecular biologists who

*Division of Biomedical & Life Sciences, Faculty of Health & Medicine, Lancaster University, Lancaster LA1 4WY, UK.

are finding it difficult to obtain research funding for single-gene-focused projects in an age of increasingly 'big data' systems biology. In any era, it seems as if a young scientist is unlikely to retire (assuming she survives in the profession to retirement) in the same field in which she began.

Works on systems biology often begin by making some startling claims for its novelty or importance. For instance, one of the commonest generalisations concerning systems biology is that, in the words of the welcome message to the 11th International Systems Biology Conference in Edinburgh in October 2010, it 'takes a *holistic* view on biology and aims at elucidating design principles of whole biological systems rather than of individual biomolecules or single events' (italics added). Even more radically, it is sometimes stated that systems biology is a *paradigm shift*, nothing short of a fundamentally new way of doing biological sciences.[1]

Certainly, systems biology makes use of a whole raft of new technologies that matured around the millennium and in the decade that followed. Deep sequencing and other high throughput analysis tools spawned a gaggle of data cataloguing capabilities with names all ending in 'omics'. Genomics, metagenomics, transcriptomics, proteomics, lipidomics and metabolomics, to name just a few, produced data on a scale previously unimaginable in biology. Crucially, this expansion of the traditional molecular biology laboratory into a data generating factory coincided with an explosion in the power and availability of computers. Indeed, omics disciplines would scarcely be possible without some way of handling their often terabyte-sized outputs. With the power to describe whole systems of biomolecules, whole cells, and even whole organisms, at molecular levels of detail, systems biology became an inevitability.

'Order and Progress': Auguste Comte's Positivism and Its Legacy

Any discipline so intent on wholes might facetiously be described as 'wholist', but does that necessarily imply a genuine *holism*? Although the disruptive technologies of the omics revolution have transformed the practice of biology research, shedding the reductionist legacy of mid-to-late 20th century molecular biology has been difficult.[2] This is scarcely surprising once one considers exactly how deep its roots are, extending back to the *positivism* of the 19th century French visionary Auguste Comte (1798–1857), which by his death had even acquired rituals and a priesthood, the *Religion de l'Humanité*.[3] This cult aspect of positivism was briefly quite successful, especially in Brazil, and its motto, 'Order and Progress', can still be found on the Brazilian flag. Positivism also acquired political ambitions, in which the bizarre idea of European unity was stressed. Comte's proposed 'Great Western Republic' would include France, the British Isles, Germany, the Low Countries, the Iberian peninsula and Italy and would have its capital, naturally, in Paris. If that were not bad enough, Greece and Poland would also be invited to join in a second phase of 'accessory members'.[4]

Auguste Comte had a rather unhappy personal life, afflicted with mental illness, unrequited love and at least one unsuccessful suicide attempt. In the words of one unsympathetic modern critic: 'Comte was a strange individual. Indeed it would not be stretching language to say he was mad'[5] (p. 44). Despite his prickly personality and long-winded prose, or perhaps even because of it, Comte possessed a remarkable ability to influence even those who disliked him personally or had philosophical reservations about the more overarching aspects of his creed, and positivist ideas spread far beyond his narrow circle of devotees within the *Religion de l'Humanité*.[3] Even in the 21st century it is common to hear scientists, or even the general public, use positivist language, though the vast majority of them have never heard of Comte.

Comte saw all of human thought as classifiable into three modes: Theological, Metaphysical and Positive, and divided up human history on that basis. The thing that characterised science, setting it apart from the religion of the Theologians and the creative philosophising of the Metaphysicians, was that it was based solely on tangible, demonstrable, common sense evidence, in other words on what could be *positively* known —

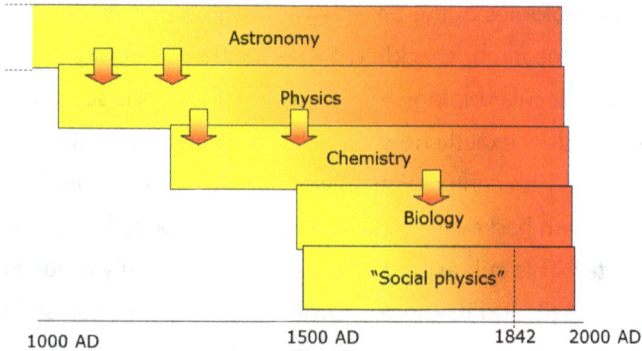

Figure 3.1. Comte's Law of Filiation of the Sciences.
Each discipline passes progressively through Theological and Metaphysical phases (yellow) before entering the Positive phase (red), at which point it becomes a true science. Comte was vague concerning the exact date of these transitions. Only 1842, the date of the completion of his major work, Cours de philosophie positive is firmly given as the year in which 'social physics' achieves the Positive phase. The transitions are aided by input from the previous discipline in the filial chain.

science was Positive. However, different sciences were at different stages of the Metaphysical–Positive transition (Fig. 3.1), and this ordering was the basis for Comte's 'law of filiation of the sciences': 'Thus we have before us Five fundamental Sciences in successive dependence, — Astronomy, Physics, Chemistry, Physiology, and finally Social Physics'[6] (p. 28) (irregular capitalisation and punctuation in original), and 'every science is [rooted] in the one which precedes it'. Each successive science had sprung forward from the previous member of the chain, with its transition into Positivity building on the established successes of its predecessor disciplines. Comte considered physics and chemistry as having achieved, by the mid-19th century, the full Positive stage of development, and biology as being nearly there. Sociology was considered to be still wallowing in the Metaphysical morass, and Comte saw it as his own specific scientific task to bring it forward into the Positive phase.

    Comte's eclectic system has often been portrayed by historians as a response to the chaos of the French Revolution and the reactionary regimes that followed it, attempting to restore order and progress, as its motto declared, to a ravaged and disillusioned France. In the following century, this spirit was reawakened in the ruins of the equally devastated Hapsburg Empire in central Europe and, amid the cosmopolitan *Kaffeehaus* culture of Vienna, positivism became *logical positivism*.

The Vienna Circle and Classical Reductionism

The Vienna Circle was formally instituted as the Ernst Mach Society on 23rd November 1928.[7] It managed to clean up Comte's positivism, stripping away the religious and political accretions and creating a version of refined purity, and if ever there was a philosophy suited to those of a purist inclination, it is the *logical positivism* of the Vienna Circle. Moritz Schlick (1882–1936) and his Vienna Circle colleagues recast Comte's concept of the unique value of *that which could be positively known*, as a means for creating a boundary criterion between the meaningful and the meaningless. In the new logical positivism, theological and metaphysical statements were not wrong, but merely senseless. The most charitable thing that could be said for them was that they perhaps had some subjective artistic validity, comparable with the meaning to be found in music or literature. The physicist Ernst Mach (1838–1916), after whom they took their official name, had pioneered an extreme form of this neo-positivist attitude in his rejection of the reality of common physical concepts such as atoms, relegating them to the dustbin of metaphysical constructs.[8] In the words of Vienna Circle member Philipp Frank (1884–1966): 'physics is nothing but a collection of statements about the connections among sense perceptions, and theories are nothing but economical means of expression for summarising these conditions'[9] (p. 220).

Physics, freed from all metaphysical trappings, was the natural foundation stone upon which the rest of science could be constructed. The resulting hierarchy therefore repeated Comte's law of filiation, in essentially the same order but rotating Comte's linear succession into a stacked structure (Fig. 3.2). Chemistry, depending as it does on the atomic laws of the physicist, is the next level in the structure, sitting on top of physics. This is followed by biology,

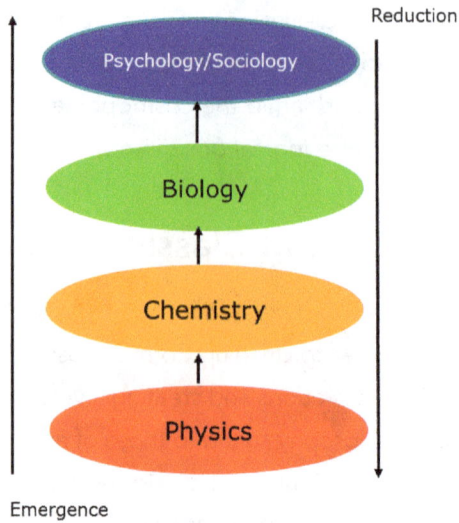

Figure 3.2. The layered mode of the reductionist hierarchy.
Statements in each discipline are re-expressible in terms of, or reducible to, statements in the discipline immediately below it.

or possibly biochemistry, then cell biology, physiology and finally things like psychology and the social sciences. This *layered model* is so ingrained into our current view of science that it is a little odd to imagine that it is barely 90 years old. The logical positivists' achievement was to begin with a difficult philosophical concept wrung from complex wrangling about meaning and evidence, and turn it into a framework for the explanation of one scientific discipline in terms of another: *inter-theoretic reduction*. By means of the process of inter-theoretic reduction '*the whole of Science becomes Physics* ... every scientific statement can be interpreted, in principle, as a physical statement'[10] (pp. 98–99) (capitals and italics in original), or as Thomas Nagel has expressed it more recently: 'Reductionism is the idea … that physics is the theory of everything'[11] (p. 3). Scientific disciplines are mere flags of convenience within a single physics-based *Unified Science*.[8]

The Vienna Circle never quite achieved its aims in full. Its leader, Moritz Schlick, was murdered by a student in 1936 and the *Anschluss* of Austria in 1938 sent many of its main members into exile. However, just in time, the ideas of the Vienna Circle had entered the English-speaking world through the publication of A. J. Ayer's (1910–1989) *Language, Truth and Logic* in 1936, probably the nearest thing to a bestseller that philosophy has ever seen.[12] It is therefore unsurprising that in molecular biology, Francis Crick's (1916–2004) reductionism seems to be sung straight from the logical positivist hymn sheet: 'The ultimate aim of the modern movement in biology is in fact to explain *all* biology in terms of physics and chemistry' (italics in original)[13] (p. 10).

Reductionism has achieved possibly its most extreme form in *singularitarianism*. Originating in the work of Raymond Kurzweil[14] and having some affinities with Frank Tipler's 'omega point',[15,16] the singularity is a future date at which Moore's Law[17] on the exponential growth of processing power has produced computers of such power that everything can be computed and we will therefore know everything. Kurzweil believes this point will be reached as soon as 2045. Even sooner than that, he claims, computers will be capable of modelling our own cognitive functions, and therefore consciousness, so accurately that we could upload copies of ourselves *in silico* and live immortally in cyberspace. An exact copy of our brain structure, down to the atomic level of every neuron would, the singularitarians believe, exhibit the same thoughts as the real thing, the same emotions, tastes and memories. Its bodily substance would be metal, plastic and silicon chips rather than proteins, lipids and carbohydrates, but those copies would nevertheless be *us* and our disembodied selves would feel our existence as being in the machine — or perhaps spread over several machines in a computing cloud.

Kurzweil's thesis has enormous emotional appeal, promising that all the world's problems will be solved, even our own individual mortalities indefinitely postponed, as long as the inexorable march of Moore's Law continues. But it also requires that reductionism be correct. All biology has to be physics, and all problems have to be computational problems, for the singularitarian vision to be achievable. The world must be merely a sum of atoms, and our understanding of the world no more than a sum of bytes and bits, or else it will fail. Even if this were true, however, there is another problem with the singularitarian project.

Even relatively trivial brute force calculations on computers can require exponentially increasing processing times. Bremermann's Limit,[18] the theoretical absolute maximum processing speed, at which every atom of the computer is vaporised as its mass is entirely converted into energy, is insufficient to generate answers to some basic combinatorial problems.[19] Reductionism, and especially its extreme singularitarian variant, breaks down on its epistemology. Even assuming we can solve every scientific problem just by computing it, we would need to wait forever to do so.

From Classical Holism to Neo-Holism

Around the time that the Ernst Mach Society was organising its first formal meetings in Vienna, Jan Smuts (1870–1950) was taking a break after his first stint as South African Prime Minister, to write *Holism and Evolution*,[20] coining the word holism from the Greek ὅλος (a whole). Like Comte, Smuts had great ambitions for his philosophy: 'All the problems of the universe, not only those of matter and life, but also and especially those of mind and personality, which determine human nature and destiny, can in the last resort only be resolved — in so far as they are humanly soluble — by reference to the fundamental concept of Holism'. For Smuts, the *vera causa*, an innate tendency for stable wholes to form from parts, occurring at all levels from the atomic through the biological to the psychological, steered the entire universe. The original conception of holism was more metaphysical than scientific, and Smuts did not acquire the same cult following as Comte had done a century previously. Nevertheless, the term holism moved into the world of science and began to be used more generally by opponents of reductionism.

One of the most intriguing critiques of reductionism in biology was supplied by Walter Elsasser (1904–1991), a quantum and geophysicist who, while working in Paris in the 1930s, had been inspired to think about biology by the physiologist Théophile Kahn (1896–1986). Elsasser was by no means the only holist biologist of the post-war era — the names of

Paul Weiss (1898–1989) and Conrad Waddington (1905–1975) are often mentioned in this context[21,22] — but Elsasser's holistic vision was more fundamental than any of his contemporaries. Elsasser's biology came to be characterised by a wholesale rejection of the reductionist model of the Vienna Circle as implemented in molecular biology. What makes Elsasser's holism a *neo*-holism rather than a successor to that of Smuts, was that he insisted that it be based on the most fundamental of physical theories, quantum mechanics, and that he rather curiously still described himself as a positivist[23] (p. 33).

Elsasser only began to publish in biology in the late 1950s, by which time he had decamped to the USA, after some two decades of digesting Théophile Kahn's ideas. Elaboration of his critique of reductionism was to occupy him for most of the 1960s and into the early 1970s. Elsasser does not merely attack the mainstream biological reductionism of the kind popularised by Francis Crick but, rather more radically, attacks the whole notion of molecular determinism in biology, using a difficult argument he named *the Principle of Finite Classes*. Elaboration of this argument would require a chapter in its own right, and has been done elsewhere,[24] but in essence it argues that wave function collapse, the phenomenon that produces the deterministic world of observable phenomena from the indeterminate world of quantum mechanics, only applies to simple molecules, such as those studied in the physics and chemistry laboratory. The complex molecules of biology — things like proteins, carbohydrates, lipids and nucleic acids — do not achieve wave function collapse, and therefore are always liable to behave in an indeterminate manner[25,26] (p. 169; p. 286). For Elsasser, much of biology was in fact 'acausal'.[27] Figure 3.3 summarises the argument in graphical form.

Elsasser's argument against determinism in biology stimulated some inconclusive critiques in the 1960s, and then faded into obscurity, having been neither conclusively disproved nor having found many adherents. It was, perhaps, a casualty of its own difficulty — few can feel equally comfortable in both the fields of quantum mechanics and biology — and

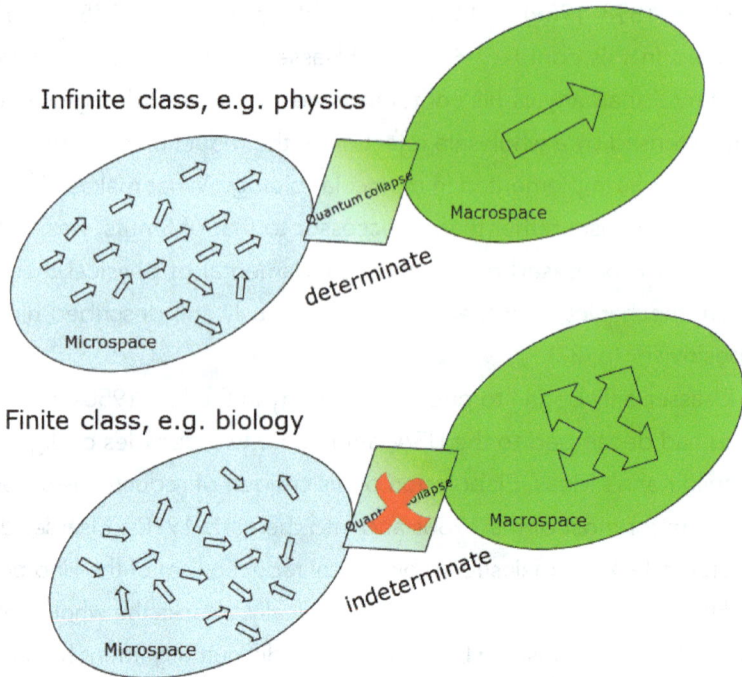

Figure 3.3. Elsasser's Principle of Finite Classes argument.
In a simple physical system — an infinite class — there are sufficiently few quantum states (small arrows) that they can be averaged (large arrow). In a complex biological system — a finite class — there are so many that no average may be obtained (4-headed 'arrow'). There is therefore no causal connection between the microscopic and macroscopic worlds for biological objects.

Elsasser's own apparent reluctance to engage directly with his critics. Indeed, by 1969, when Elsasser declared that he was 'therefore addressing the present scheme mainly to younger people whose philosophy may not yet have approached a point of condensation'[27] (p. 503), one can almost hear the electric guitars wailing in the background.

Nevertheless, despite Elsasser's inability, or reluctance, to force his theory into the mainstream, his influence on modern neo-holism remains profound, because even if his anti-deterministic argument was flawed, it produced, as a by-product, an alternative to the layered model of Vienna Circle reductionism. For Elsasser, biology was the science of the complex, and therefore is a superset of all the other sciences which deal with subject matter of greater regularity than the messy stuff of biology. Chemistry and physics are subsets of 'biology' (as Elsasser conceived it), activities that commence when we start to refine our area of study down to the molecular and atomic level — they are simply the areas of 'biology' where determinism and causality apply (Fig. 3.4).

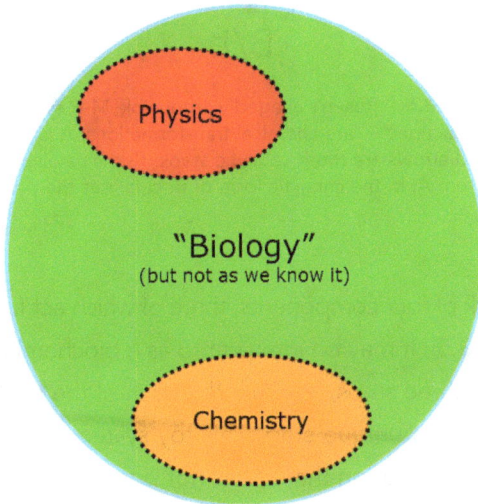

Figure 3.4. Walter Elsasser's alternative to the layered model (see Fig. 3.2). Rather than biology emerging from chemistry or physics, the latter are sub-fields of a new science of complex systems. Physics and chemistry are not more fundamental than biology, but are actually rather specialist areas of biology which deal with infinite classes, i.e. with simple, homogeneous subject matter. See also Fig. 3.3.

[M,R] — The Irreducible Paradigm of Relational Biology

Elsasser's influence channels into modern holism through Robert Rosen (1934–1998), whose mathematical work, which he collectively termed *relational biology*, has achieved the status of a Mrs Rochester in systems biology's attic. The centrepiece of Rosen's critique of reductionism is the *[M,R] system* — standing for Metabolism/Repair. [M,R] is a small self-

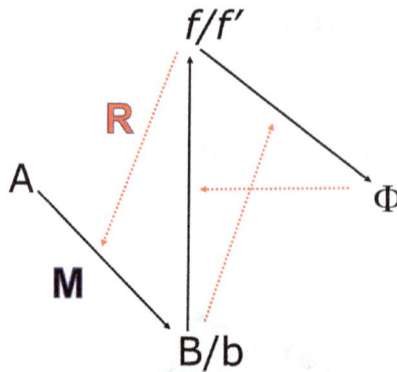

Figure 3.5.   Rosen's allegedly irreducible M–R system.
Full arrows — the M reactions — are chemical transformations: A is converted to B, B to *f*
and *f* to Φ. Red dotted arrows are the R catalytic steps: *f'*, the catalytic form of *f*, catalyses
the production of B from A, b, the catalytic form of B, catalyses the production of Φ from *f*
and so on.

referential network of four components, three of which act functionally within
the network (Fig. 3.5). It may be interpreted as a biochemical pathway with
four moieties and three steps, in which the each of the final three moieties
are also catalysts for a unique step. This toy system was the subject of a
mathematical demonstration that it was not possible to predict the properties of its entirety through an analysis of the properties of its individual
components.[28,29] [M,R] is therefore not reducible to its component parts
and can only be understood as a whole. Rosen took pains to give [M,R] as
few parts and functions as possible — it is the self-referential nature of its
structure, the way that three of the four components are necessary for the
production of three other components of the same set of four, that causes
the breakdown in the reductionist hierarchy. Irreducible complexity does
not require a big and complicated system, but can be present in tiny toy
systems like [M,R]. Indeed, Rosen defines complex systems as those which
cannot be reduced.

With Rosen, we are no longer in the business of the epistemological anti-reductionism which defeats the singularitarian argument. Rosen's
anti-reductionism was *explanatory*. Irreducibility is a property independent

of the size of the thing that cannot be reduced. [M,R], he claims, simply cannot be explained using a reductionist approach. Epistemological anti-reductionism is a *holism in practice*, an observation that certain components of the reductionist programme are infeasible. Explanatory anti-reductionism is a *holism in principle*.

Rosen draws comparisons with Elsasser's nested model — as a complex system [M,R] as a whole is situated in the domain of 'biology' or the science of complexity whatever name one gives to it. Break [M,R] down into its component individual steps and these are then in the domains of 'physics' or 'chemistry'. However, what we can say mathematically about the behaviour of the component parts, cannot be subjected to any additive process that will allow us a complete mathematical description of the whole. Thus what we can say in the molecular biology laboratory about single proteins or genes does not tell us, *contra* Crick, all that we would wish to know about the whole organism from which they are isolated, even if we have full knowledge of all of the components and how they function in isolation.

Rosen also ventured that [M,R] could have a more general interpretation. As well as representing a single self-referential network, the component parts could be taken to represent sets of reactions in living things. For instance, the first step could represent not just one, but all, metabolic reactions, the second and third steps sets of other reactions necessary to ensure that metabolism can continue — hence the Metabolism/Repair name. Pursuing this set-oriented interpretation of [M,R], Aloisius Louie has described its components in terms of formal set algebra.[30] Louie's central result from this analysis is the identification the presence of an impredicative set within [M,R], meaning a set that is member of itself. Impredicative sets are non-computable on a Turing machine, which remains the basic conceptual architecture of all computers. By implication, no complex system and no biological system can therefore be fully functionally modelled *in silico*.

Biology, if Rosen and Louie are correct, is therefore not only non-reducible in the laboratory but also cannot be modelled on a computer.

The seriousness of this conclusion for systems biology as it is currently practised, has generated a stunned silence punctuated by occasional attempts at refutation. However, the various attempted disproofs[31,32] of Louie and Rosen's work have also proved technically controversial, and the resulting lack of clarity has not served the debate well.[33] Rosen's relational biology has achieved a higher profile than Elsasser's work, insofar as systems biologists are often aware of its existence,[34,35] but its technical difficulty for those without the required background has left the adjudication on its validity to a small number of jurors who cannot reach a unanimous verdict.

Neo-Reductionism: Software Laws, Physics and Stamp Collecting

Anyone who has been an undergraduate in genetics or molecular biology since the 1970s will be familiar with the workhorse examination question, 'What is a gene?' Generations of students are thereby invited to do a little inter-theoretic reduction in the spirit of the Vienna Circle, expressing the higher-level abstract explanations of genetics in terms of the nuts and bolts of molecular biology. Prior to its arrival in the examinations hall, this topic had formed the basis of Kenneth Schaffner and David Hull's (1935–2010) attempts, in the late 1960s and early 1970s, to apply Vienna Circle reductionism to biology.[36,37] It soon became obvious to them that this was not going to be easy.

There is, for instance, no term in molecular biology that can capture everything that is implied by the term *gene* in classical genetics. Molecular biologists know that genes are made of DNA, but each gene is unique in terms of how that DNA is constituted into that particular gene. To reduce processes, such as segregation or gene silencing, the difficulties are even greater. For the process of meiosis — the independent assortment of genes during gamete formation — the abstract genetic explanation is both far easier and far more illuminating than any attempt to be more specific about molecules. Indeed, meiosis can be represented simply in terms of a set of rules for moving objects. Even if those objects are not actual chromosomes but simulations, e.g. beanbags or graphic objects in a computer simulation, the same rules would apply, and that DNA-free explanation would be a fully adequate one.[38] Even if there were no actual biological objects, the theory would still make *logical* sense.

In a systems biology context, one might derive novel rules concerning a set of properties of a gene-regulatory or metabolic network. These rules might turn out to have logical validity in other contexts and different kinds of network, perhaps even in non-biological networks. Of course in the real biological world, Mendelian and Darwinian and metabolic systems phenomena are instantiated in DNA, cells and organisms, but the laws we use to describe their behaviour are often independent of their substrate, what Paul Davies has called *software laws* rather than *hardware laws*.[39] Reductionism does not so much fail here as appear to be an unnecessary complication. Schaffner therefore replaced classic inter-theoretic reduction with a more pragmatic *biological principle of reduction*,[36] a commitment to try to reduce wherever possible, to create as many reductive explanations as possible, even in the absence of complete reduction.

Although this softened Crick's vision of explaining *all* of biology in terms of physics and chemistry, it still placed biology within the layered model of the Vienna Circle. The next stage in the evolution of reductionism, taken by Alexander Rosenberg, was to replace that model.[40,41] Rosenberg proposes in its place a two-layer model (Fig. 3.6) building on the work[42] of J. J. C. Smart (1920–2012). The lower layer is physics and the upper layer is termed 'engineering'. This is not to be interpreted literally, but to serve as a shorthand for all sciences other than physics. An 'engineering' question is one that does not require an answer that includes a full physical explanation, but one for which chemistry, biology, psychology, etc. will suffice. Ernest Rutherford (1871–1937) previously made a similar analogy, but replaced 'engineering' with 'stamp collecting'.

Neo-reductionism proposes that physics provides the description of the molecular order of a system — its *micro-state* — whereas 'engineering'

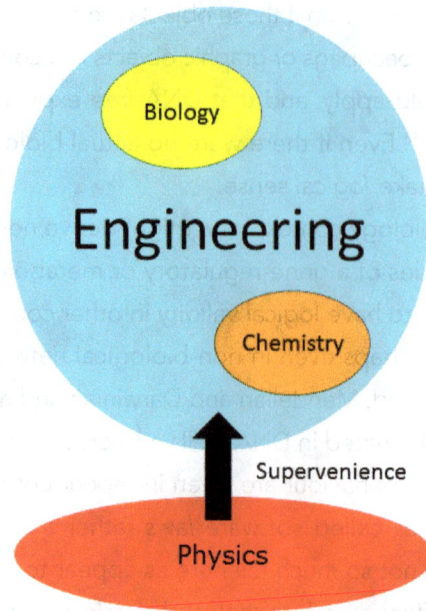

Figure 3.6. Neo-reductionism's simplification of the layered model. Neo-reductionism's flat model, an alternative hierarchy to Fig. 1, proposed by J. J. C. Smart but previously implied by Ernest Rutherford: 'either physics or stamp collecting'.

explanations refer to supra-molecular configurations of that system — its *macro-state*. All macro-states are *supervenient* on underlying micro-states. Supervenience implies that a given micro-state will always result in the same macro-state. By contrast, a macro-state may have more than one micro-state that will give rise to it (Fig. 3.7). Micro-states therefore determine macro-states, but macro-states are not reducible to micro-states, or at least not reducible to unique micro-states. Davies' software laws are therefore laws of the macro-state and not reducible to those of the micro-state. The software laws, however, cannot allow behaviour which breaks the laws of the micro-state.

Moving outwards beyond individual systems, neo-reductionism sees the universe as consisting of a *micro-space*, its objective atomic/quantum physical reality, and a *macro-space*, which is the configuration of larger

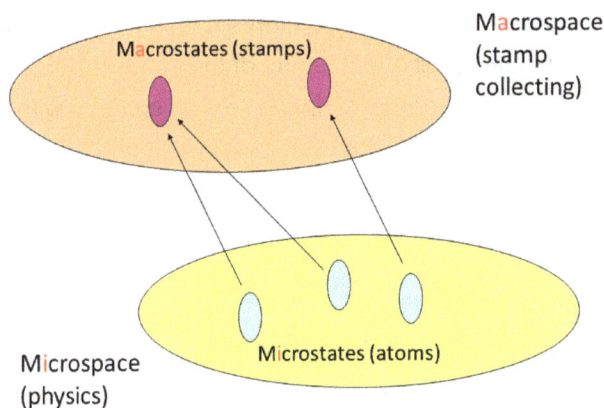

Figure 3.7. The non-unique dependence of macro-states on micro-states. Classical reductionism has always struggled to determine how macro-space configurations are determined by their underlying micro-spaces, in other words how to model a macro-space. Neo-reductionism implies that it is more important to understand how micro-space variation affects that of macro-space, in other words how macro-space is realised.

order entities studied by all the other sciences, and which is supervenient on the micro-space. Interestingly, Elsasser had already outlined a similar concept as part of his anti-reductionist argument. His concept of *variostability* refers to the tendency of a macro-state of a system to remain coherent in the presence of micro-state variation.[43] Although Elsasser argued that this had not merely anti-reductionist but even anti-determinist implications, his conclusions are dependent on the assumption that reductionism requires a unique micro-state to macro-state binary mapping. Neo-reductionism of the kind advanced by Rosenberg, however, allows for a one-to-many mapping between macro-state and micro-states. This is compatible with variostability.

Another parallel with the work of Elsasser is his concept of *biotonic laws*, which he hypothesised as laws pertaining solely to biological systems, which were not reducible to underlying physical laws.[23] Davies' software laws concept can be made to fit partially with the biotonic laws concept, just as variostability can be made partially congruent with supervenience. Neo-holism and neo-reductionism thus begin to find common ground. Lest we become too enthused over the prospects of synthesis, however, it should be borne in mind that Elsasser drew his conclusions from his controversial quantum mechanical theory of the Principle of Finite Classes, which implied total indeterminacy for biological systems with respect to their physical constitution. Neo-reductionism's concept of supervenience requires a causal relation pointing upwards from micro-state to macro-state and therefore is deterministic in a way that Elsasser would have rejected.

Modelling versus Realisation

Inter-theoretic reduction requires explanations of how one theory can be expressed in terms of a lower level theory in the layered hierarchy. So the macro-space configurations of one discipline must be shown to be deter- mined by the underlying micro-spaces of the theory below. In other words, reductionism requires us to be able to model macro-spaces in terms of a micro-space. Robert Rosen had stern words for reductionists on the sub- ject of modelling, making a distinction between a true model and a mere simulation. Imagine an attempt to build a piece of software to represent a biochemical network. No matter how accurate the representation of the entities of the system may be, they are only in a model *sensu stricto* if the rules that connect them and govern their behaviour are also accurate rep- resentations of the laws of the natural world. Otherwise the software is a simulation. The *entailment structure* — the framework of rules that govern how bits of the system interact — of a true model faithfully represents the corresponding entailments of the thing being modelled.[28] Failure to do so will generate a *black box*, a simulation which may be very good at pre- dicting the output given a set of inputs, but which does not represent any true understanding of the system. Simulations are merely *ad hoc* black box predictors of the phenomenological behaviour of whatever they simulate.

This is indeed a tough requirement to satisfy in any field. Simulation, rather than modelling in Rosen's sense, is the norm in most cases. Rosen might be implying that when reductionists often think they have achieved inter-theoretic reduction, the reality may be considerably less conclusive.

However, Schaffner's biological principle of reduction acknowledges that this is often the case, at least in biology if not other disciplines, and accurate, if black box, simulation may be a good preparation for the deduction of a true model of the actual entailment structures of the system concerned.

Even assuming that both aspects of an inter-theoretic reduction were completed to Rosen's satisfaction, with adequate re-expression of both entities and entailment structures, the completed model of reduction is still vulnerable to change in either, or both, of the reduced and reducing theories. If some aspect of biology is regarded as reduced to physics or at least chemistry, the assumption is made that the underlying chemistry is correct. Should the relevant parts of chemistry be disproved, the reductional chain will be broken, and the biology will require to be re-reduced to the new physics or chemistry. Previous satisfactory reductions may suddenly become invalid in this way, and reduction must always therefore be considered to be provisional. However, just as a previous reductive chain may be broken by changes in the underlying theory, so may reduction become possible where before it was not.[44,45] Anti-reductionist declarations must always be provisional too. Ernest Nagel (1901–1985)[46] points out that chemistry is only reducible to post-1925 physics, and thermodynamics is only reducible to post-1866 statistical mechanics. In both cases it was advances in the lower theory (physics) that enabled the reduction, not any new insight into chemistry or any improvement in reductionist method. The emerging consensus from the mid-1920s onwards that chemistry was finally reducible to physics, was one of the factors that spurred the enthusiasm of the Vienna Circle to apply reductionism everywhere.

Neo-reductionism avoids Rosen's strictures regarding modelling and simulation, not by merely adopting the biological principle of reduction and/or conceding that reductions can remain provisional, but by replacing modelling with *realisation*. Neo-reductionism's concept of a non-unique macro-space to micro-space supervenience relationship implies that it is more important to understand how micro-space variation affects that of macro-space than to model the macro-space on the micro-space. In other words, the question becomes how a macro-space is *realised* upwards from its underlying micro-space, rather than how to reduce/model downwards from macro-space to micro-space.

This notion has some considerable implications for scientific method. The inversion required by neo-reductionism would seem to require a kind of *Gestalt* switch in the way that our brains do science. A plan for what science would look like once we have managed to perform that change in perspective is not obvious. Nevertheless, some intriguing hints are visible in the work of Stephen Wolfram on *cellular automata*. An automaton is a software entity that performs certain behaviours under a simple set of rules.[47] These may be very clear and straightforward — cellular automata have none of the knotted puzzle nature of [M,R]. However, they may exhibit remarkably rich patterns of activity. Conway's Game of Life is the most famous example of a cellular automaton, where different starting configurations result in radically different shifting outcomes. Crucially, we know that cellular automata are deterministic — we programmers have specified their rules and they always abide by them. We can therefore say that we fully understand the micro-space of the automaton. Wolfram produces a vast variety of such automata showing how complexity can arise from simple starting conditions and conversely how order can emerge from chaotic conditions. According to Wolfram, there are many instances of automaton-like behaviour across a wide range of systems from physics to psychology. Modelling a system is therefore less important than the question of whether the system behaves similarly to a known automaton, i.e. realisation.

For neo-reductionism, systems biology is labouring under the weight of the modelling problem, whereas it ought to be recasting itself in terms of the realisation problem. The modelling problem founders both on the sheer scale of the data — it cannot counter epistemological anti-reductionist arguments — and also potentially on the hidden problems of self-referential systems — if Rosen is correct, it cannot counter explanatory anti-reductionist arguments either. Even leaving Rosen aside, it is evident that the modelling problem is the problem of data increasing faster than the conclusions we can draw from it — a problem of where to end. The realisation problem, by contrast, is one of where to start.

The Scientific Understanding of Science

The reductionism–holism debate in systems biology sits within a wider context that goes beyond the bounds of daily activity in the research lab. This anthology is on the subject of the public understanding of biology. The public understanding of science in general has become a speciality in its own right with endowed chairs in prestigious universities and 'public engagement' high on the priority list of league table-driven British universities. However, this phrase implies that a failure to understand science is largely something 'out there' in the public. If only scientists can learn to communicate better, the public will understand better.

Valuable as such evangelical work is — and after all, it is the public who are paying the major portion of the salaries of the scientists, so they are entitled to know where their money is going — it misses a problem rather closer to home: scientists often do not have a very firm conception of their own working methodology, and even less of a comprehension of the methodology of other scientific disciplines. If anybody is looking for another chair to endow, a Professorship in the Scientific Understanding of Science would be both a provocative and valuable contribution.

The origins of this problem lie in the fact that most undergraduate science courses do not teach the philosophy of science. If they do, all that will be included will be an exhortation to perform experiments with careful controls that test hypotheses and seek to falsify rather than confirm them — in other words, most biologists, if they think about scientific method at all, are Popperians.[48] Part of this is due to the advocacy of Karl Popper's (1902–1994) legacy by Richard Dawkins and his predecessor as the UK's favourite popular writer on biology, Peter Medawar (1915–1987), whose best-selling *Advice to a Young Scientist* laid down the Popperian law to many aspiring young molecular biologists of the 1980s and beyond.[49]

However, two things are rarely if ever mentioned when Popper is discussed by biologists: his anti-reductionism, which became stronger in his later years,[50] and the apparent anti-Darwinism of his late period (although his exact stance on this is still a matter of controversy). Biologists in the 21st century thus usually hold an incompatible mixture of philosophical views on their own subject — a classical reductionism channelled from the Vienna Circle via A. J. Ayer to Francis Crick, rubbing shoulders with the post-Vienna Circle thought of Popper. Often these are held at such an unconscious level that biologists will deny having any philosophical thoughts at all, believing all such things to be irrelevant to science.

Scientists are therefore in a poor position to defend their discipline against those who would cast doubt on its entire existence. To take a few common examples, sociologists of the 'science studies' or 'science, technology and society (STS)' persuasions seek to represent science as a set of rituals performed by a secular priesthood. Neo-Marxists see it as a bourgeois activity devoted to replicating the existing political structure, and have a particular antipathy to biology as an obstacle to their notions of the infinite malleability of the human social order. Social constructionists wish to deny any discipline that believes an objective view of reality can be achieved. These are caricatures, of course, necessitated by brevity, but these threats to science in its current form are real. In order to defend ourselves, scientists need a clearer idea of who we are, what we are doing and why, that goes deeper than the currently fashionable notions of 'impact' and 'engagement'. Part of the formation of that clearer idea must come from a deeper understanding of our philosophical underpinnings. Systems biology has an opportunity to lead the way in this endeavour, given that the field sits on the cusp of a profound philosophical decision about its future orientation. Systems biology may make scientists become natural philosophers once more.

References

1. Marcum JA. (2008) Does systems biology represent a Kuhnian paradigm shift? *New Phytol* **179**(3): 587–589.
2. Gatherer D. (2010) So what do we really mean when we say that systems biology is holistic? *BMC Syst Biol* **4**(1): 22.
3. Simon W. (1963) *European Positivism in the Nineteenth Century. An Essay in Intellectual History.* Cornell University Press, Ithaca.
4. Comte A. (1880) *A General View of Positivism, 2nd ed.* Reeves and Turner [1851], London.
5. Wilson A. (2000) *God's Funeral.* Abacus, London.
6. Comte A. (1853) *The Positive Philosophy of Auguste Comte.* John Chapman, London.
7. Stadler F. (1992) The 'Verein Ernst Mach' — what was it really? In: Blackmore J (ed.) *Ernst Mach — A Deeper Look.* Boston Studies in the Philosophy of Science, 143, pp. 363–78. Kluwer, Dordrecht.
8. Frank P. (1970) Ernst Mach and the unity of science. In: Cohen RS and Seeger RJ. (eds.) *Ernst Mach Physicist and Philosopher.* Boston Studies in the Philosophy of Science, 6, pp. 235–44. D. Reidel, Dordrecht.
9. Frank P. (1970) The importance of Ernst Mach's philosophy of science for our times. In: Cohen RS and Seeger RJ (eds.) *Ernst Mach Physicist and Philosopher*, pp. 219–34. D. Reidel, Dordrecht.
10. Carnap R. (1995) *The Unity of Science.* Thoemmes Press, Bristol.
11. Nagel T. (1998) Reductionism and antireductionism. In: Bock GR and Goode JA (eds.) *The Limits of Reductionism in Biology.* Novartis Foundation Symposium, 213, pp. 3–14. John Wiley and Sons, Chichester.
12. Ayer AJ. (1936) *Language, Truth and Logic.* V. Gollancz, London.
13. Crick F. (1966) *Of Molecules and Men.* University of Washington Press, Seattle.
14. Kurzweil R. (2005) *The Singularity is Near. When Humans Transcend Biology.* Viking/Penguin, London.
15. Tipler FJ. (1994) *The Physics of Immortality: Modern Cosmology, God and the Resurrection of the Dead.* Doubleday, New York.
16. Tipler FJ and Barrow JD. (1986) *The Anthropic Cosmological Principle.* Oxford University Press, Oxford, UK.
17. Moore G. (1965) Cramming more components onto integrated circuits. *Electronics Magazine.* (19 April 1965).
18. Bremermann H. (1982) Minimum energy requirements of information transfer and computing. *Int J Theor Phys* **21**: 203–217.

19. Gatherer D. (2007) Less is more: The battle of Moore's Law against Bremermann's Limit on the field of systems biology. *BMC Syst Biol* **1 supp. 1** (1 supp.1): 53.

20. Smuts JC. (1926) *Holism and Evolution*. Macmillan and Co, London.

21. Drack M and Apfalter W. (2006) *Is Paul Weiss' and Ludwig von Bertalannfy's System Thinking still Valid Today?* 50th Annual Meeting of the International Society for the Systems Sciences.

22. Gatherer D. (1996) A stroll across the epigenetic landscape: Bringing Waddington's ideas into molecular biology. *Early Pregnancy* **2**(4): 241–243.

23. Elsasser WM. (1961) Quanta and the concept of organismic law. *J Theor Biol* **1**: 27–58.

24. Gatherer D. (2008) Finite universe of discourse. The systems biology of Walter Elsasser (1904–1991). *Open Biol* **1**: 9–20.

25. Elsasser WM. (1962) Physical aspects of non-mechanistic biological theory. *J Theor Biol* **3**: 164–191.

26. Elsasser WM. (1969) The mathematical expression of generalized complementarity. *J Theor Biol* **25**(2): 276–296.

27. Elsasser WM. (1969) Acausal phenomena in physics and biology: A case for reconstruction. *Am Sci* **57**(4): 502–516.

28. Rosen R. (1991) *Life Itself: A Comprehensive Inquiry into the Nature, Origin, and Fabrication of Life*. Allen T and Roberts D (eds.) Columbia University Press, New York.

29. Rosen R. (2000) *Essays on Life Itself*. Allen T and Roberts D (eds.) Columbia University Press, New York.

30. Louie AH. (2009) *More than Life Itself. A Synthetic Continuation in Relational Biology*. Ontos Verlag, Frankfurt.

31. Chu D and Ho WK. (2006) A category theoretical argument against the possibility of artificial life: Robert Rosen's central proof revisited. *Artif Life* **12**: 117–134.

32. Landauer C and Bellman K. (2002) Theoretical biology: Organisms and mechanisms. *AIP Conf Proc* **627**(1): 59–70.

33. Louie AH. (2011) Essays on more than life itself. *Axiomathes* **21**: 473–489.

34. Cornish-Bowden. (2015) Tibor Ganti and Robert Rosen: contrasting approaches to the same problem. *J Theor Biol* **381**: 6–10.

35. Cornish-Bowden A and Cardenas ML. (2007) Organizational invariance in (M,R)-systems. *Chem Biodivers* **4**(10): 2396–2406.

36. Schaffner KF. (1976) The Watson–Crick model and reductionism. In: Grene M and Mendelsohn E (eds.), *Topics in the Philosophy of Biology*, pp. 101–27. Boston Studies in the Philosophy of Science. D. Reidel, Dordrecht.

37. Hull DL. (1974) *Philosophy of Biological Science*. Beardsley E and Beardsley M (eds.) Prentice-Hall, Englewood Cliffs, NJ.

38. Kitcher P. (1984) 1953 and all that: A tale of two sciences. *Philos Rev* **93**: 335–373.

39. Davies P. (1987) *The Cosmic Blueprint*. Heinemann, London.

40. Rosenberg A. (1985) *The Structure of Biological Science*. Cambridge University Press, Cambridge.

41. Rosenberg A. (2007) Reductionism (and antireductionism) in biology. In: Hull DL and Ruse M (eds.) *The Cambridge Companion to the Philosophy of Biology*, pp. 120–38. Cambridge University Press, Cambridge.

42. Smart JJC. (1963) *Philosophy and Scientific Realism*. Routledge & Kegan Paul, London.

43. Elsasser WM. (1998) *Reflections on a Theory of Organisms. Holism in Biology*. The Johns Hopkins University Press, Baltimore.

44. Roll-Hansen N. (1969) On the reduction of biology to physical science. *Synthese* **20**(2): 277–289.

45. Ackermann R. (1969) Mechanism, methodology, and biological theory. *Synthese* **20**(2): 219–229.

46. Nagel E. (1960) The meaning of reductionism in the natural sciences. In: Danto A and Morgenbesser S (eds.) *Philosophy of Science*, pp. 288–312. Meridian, Cleveland.

47. Wolfram S. (2002) *A New Kind of Science*. Wolfram Media, Champaign, I.

48. Popper KR. (1963) *Conjectures and Refutations*. Routledge & Kegan Paul, London.

49. Medawar PB. (1979) *Advice to a Young Scientist*. Harper & Row, New York.

50. Popper KR. (1974) Evolution and world 3. In: PA Schlipp (ed.) *The Philosophy of Karl Popper*, pp. 1048–80. Open Court; La Salle, IL.

Part Two

Rethinking Biological Concepts

4 *Homo faber*, Will, Determinism, and Heredity: From Genetics to Epigenetics

Ilya Gadjev*

Introduction

Two centuries ago, the French mathematician, astronomer and politician Pierre-Simon Laplace posited that the present state of the universe was the effect of its anterior state and the cause of the one to follow, and if a sufficiently vast intellect would know the motion and the location of all items constituting nature then this intellect, with certainty, would have the future of the Universe, just like its past, 'present to its eyes'.[1]

The view that everything can be predicted because it is determined and necessitated by identifiable pre-existing factors and conditions spread from the field of mechanics to other realms, including the life sciences. Here, the adjusted version of scientific determinism was welcomed by a stratum of already established deterministically-resonant biological notions, and an environment of favourable scientific and social dynamics. Reaching throughout biology and medicine, the occurrence and influence of deterministic thinking became particularly strong in the study of inheritance. The formation of genetics as a pre-eminent biological discipline is intertwined with ideas of

*Honorary Senior Research Associate, UCL Institute of Education, University College London, 20 Bedford Way, London WC1H 0AL, UK.

determinism and it is to a large extent through the authority of genetics that such ideas gain power and spread to colonise a wider intellectual territory.

Benefiting from helpful technical and methodological developments in the 19th and 20th centuries, while fruitfully interacting with modern thinking and the broad anxieties of the period, the genetic–deterministic approach to understanding and controlling the living world flourished and emerged as an important element not only in the sphere of science but also in the greater mosaic of ideology, politics and mass media characterising world history in the last hundred years or so.

Genetic or biological determinism is not monotypic but exists as a gradient of various types: from 'hard' *'gene for'* understandings where a gene is seen as determining a trait (no matter how complex), to more intricate, multifactorial, 'soft' notions where many genetic, cellular, developmental and other biological factors are linked to a phenotype in interplay with the conditions of the micro- or macro-environment. Just like in physics, the last decades have also seen a shift towards 'softer', more complex forms of determinism in the life sciences.

Revealing remarkable aspects of multidimensionality and interactive openness, recent developments in biology, including studies in epigenetics, have substantially enriched our views on heredity. The complexity of inheritance, the influence of the external and internal conditions, the intricacy of development and the role of behaviour have returned to the epicentre of the genetic discourse. The experimental and rhetorical emphasis has moved from the static and the simplistic towards the dynamic and the systemic. *Nurture*, it seems, has been re-paired with *nature*.

The move towards 'softer', more complex views of heredity where *the biological* is understood in probabilistic terms allowing strong, multifactorial interconnectedness with the environment and *the social*, might be interpreted as a Kuhnian paradigm shift. This is evident in the communication media where the advances in gene regulation and epigenetics are often presented as revolutionary and iconoclastic concerning 'conventional' genetics. Yet, despite their inherent potential to challenge simplistic deterministic thinking and stimulate a constructive re-estimation of the way heredity is perceived, these developments have, in reality, not been adequately internalised and successfully employed to address the reductionist–deterministic foundations of genetics. Understandings and rhetorical features of 'hard' inheritance have been recycled and absorbed into the structure of its new, 'soft' counterpart.[2] Consequently, the outlook and language of genetic determinism have not disappeared or significantly mutated but have come to coexist with not less striking narratives about the defining power of the environment and the will where convenient reductionist and deterministic explanations still play key roles.

An analysis of the conceptual developments in genetics and the network of factors involved in the formation of the conflicting views of 'hard' and 'soft' inheritance can elucidate this dichotomy and contribute to a more adequate perspective emerging from a multisided dialogue between the scientific disciplines, the humanities and the arts.

'Hard' Inheritance

Genetic determinism represents a radical view of the nature of heredity, development and life where the equilibrium in the interconnected '*genes <->genetic plasticity<->environment<->phenotype*' tetrad is drastically shifted towards the importance of the genome.[3] This doctrine apprehends the phenotype (development, shape, size, susceptibility to disease, behaviour, cognition, etc.) either entirely or largely through the perspective of the information encoded in the genes (sequences of nucleotides in the DNA molecules). Genetic–deterministic understandings cover all biological strata: from molecules (a nucleotide sequence has all the information for a protein or an RNA molecule — a gene determines the structure and function of biopolymers), via organisms (the genome contains all the necessary information for the development of an organism — the genome determines the structure and function of an organism), to populations, species, etc.

The position of genetic determinism in the academic and wider public space has fluctuated over time, becoming dominant in the 20th century. In the last few decades, however, the pre-eminence of this outlook has again started to wane. Scholars now widely agree that micro- and macro-environmental and other factors influence the structure of the genetic system, the realisation of the genetic 'potential', the formation of the phenotype and the transmission of traits. The genetic system itself has been shown to be stunningly complex, dynamic, sensitive and intricately regulated. Heredity has begun to be regarded in more probabilistic and less simplistically-deterministic terms. Nonetheless, this multifaceted, holistic picture has not become central, or even categorically present, in academia, the education system, the mass media and society at large. Instead, surprisingly simplistic deterministic concepts and rhetoric are still routinely employed in every sphere. The kaleidoscope of life on Earth remains generally perceived as pivoting around the genetic blueprint represented by molecular-digital DNA information.

The enduring gene-centric view of life was forged in the context of far-reaching social and scientific dynamics in the 19th and 20th centuries. The study of inheritance was burgeoning. Some theorists understood heredity as a type of force which was passed on to the new generation. Others thought that there was material, organic basis to hereditary continuity where transmitted hereditary matter could be formed either by particles from all tissues and organs of the body (e.g. Charles Darwin's *pangenesis*), or could be a separate reproductive entity (what August Weismann called 'germplasm' in contrast to 'soma').[4] Gregor Mendel's experiments in plant breeding had shown that some heritable traits were connected to certain factors residing in the reproductive structures which were passed on and combined in the offspring. Subsequent work supported these findings and solidified the understanding that there was a relationship between characteristics and specific discrete factors inherited from the parents. Wilhelm Johannsen at the beginning of the 20th century described the two distinct worlds of traits and hereditary factors as *phenotype* and *genotype*. *Genotype* was understood as the sum of all individual genetic factors, which Johannsen called *genes* and saw as potentialities or dispositions for specific traits transmitted to the progeny.[5]

Thus, *the biological* was separated not only into Weismann's *soma* and *germplasm* but also into *phenotype* and *genotype*; the environment, heredity and the properties of the organisms were delineated from each other; the inheritance of acquired characteristics was all but removed from the biological picture. It is important, however, to emphasise the abstract-ness of the key genetic terms introduced by Johannsen, which he refused to associate with or embed in a specific material medium or locus.[5] These concepts do bear strong deterministic resonances, but this determinism is rather broad and far from today's molecular-reductionist notions. In fact, recognising the totality of the nature of an organism, Johannsen warned against the use of phrases such as 'genes for specific characters'.[4] Nev-ertheless, some years later, Thomas Morgan's (among other researchers') work led to 'embodiment' of the genes: although still rather abstract, now they were understood as stable loci on the chromosomes, responsible for specific traits. As the gene was rapidly being clothed with biological matter, Darwinism was being stripped of its *pangenic* 'impurities' and amalgamated with this new increasingly reductionist and deterministic view of genetics to finally become rebranded as *neo-Darwinism*.

Despite this conceptual rigidisation, the first half of the 1900s wit-nessed a rich debate about the nature of heredity. Genetics and develop-ment were commonly regarded as dynamic, multifaceted systems, which were to be studied and understood in a statistical, probabilistic fashion.[6] This relative pluralism was dramatically impacted by the discovery of the double helix and the advances in molecular biology in the second half of the 20th century. The biological sciences became considerably physicalised. Yet, whereas many domains of physics had become highly probabilistic, heredity was becoming highly mechanistic and deterministic — charac-terised by molecular rigidity, stability, specificity and relatively simplistic causality.[6] To study and discuss genetics in a serious and acceptable fashion now meant to operate within the narrow epistemological and ontological confines of digitalised hereditary units seen as sequences of nucleotides in the DNA molecules. The view that the genetic information and program is contained and encoded solely in the nucleotide sequence and flows in a precise deterministic fashion to RNA and protein molecules was enshrined in James Crick's 1958 *Central Dogma of Molecular Biology*. Although the *Central Dogma* has been refined and enriched over the decades, its digital, reductionist and deterministic spirit is still fundamental to the way genetics is perceived and presented.

A graphic example of the gene-centric view of living nature, where knowing the molecular gene means knowing the phenotype it encodes and studying genetics is, in a sense, like studying the basis of life, its 'essence' is the '*gene for*' type of biological determinism. Although it is common knowledge that most characteristics are linked to many genes, which interact with each other and the environment in complex and not easy to decipher ways, the '*gene for*' statements remain widespread and emblematic for the entire biological–deterministic discourse.

Scientists, it seems from the news, are incessantly on the hunt for the one gene determining characteristics ranging from colour, smell, size, shape, mass, strength, productivity, fertility, development, health and disease; to partner matching, addiction, sexual orientation, educational achievement, behavioural patterns, and wellbeing. Particular attention has been given to 'celebrity genes' such as the '*binge-drinking gene*', the '*happiness gene*', the '*gene for speed*', the '*bargain-hunting gene*', the '*gay gene*', the '*infidelity gene*' and, among many others, VMAT2, the '*God gene*'.[7-13]

Clearly, the transition towards stricter ontologisation–essentialisation of the gene and the popular success of molecularised, deterministic genetic thinking cannot be attributed only to advances in science and technology but are also inseparable from the shaping influence of beliefs, archetypes, hopes, fears, and other aspects of the human condition. The predictable, sequenceable and controllable reductionist–deterministic gene has become an epistemic multi-tool not only conveniently applied to study and explain biological inheritance and the nature of life but also systematically involved in technocratic visions of human protection and betterment interlinked with political ideologies and schemes for monetary gain.[14]

A telling illustration of the issues surrounding the molecular-geneticisation of the human situation, and particularly, complex features like character and behaviour is the story of *MAO*, the 'warrior' or 'criminal gene'. *MAO-A* encodes a specific form of an enzyme involved in the degradation of important neurotransmitters like dopamine, adrenaline and serotonin. A 1990s study on the 'abnormal behaviour' of male members from a large Dutch family concluded that their impulsive-aggressiveness was due to a mutation in this gene, leading to MAO-A deficiency.[15] This work became axiomatic to the discussion that genes rather than choice or decision-making are central to crime. Arguments emphasising genes over choice have since been echoing from courtroom to classroom and from media to academia across the globe.

The journal *Nature* reports that since the introduction of *MAO-A* testing to the judicial system in 1994, there have been hundreds of cases in the United States and more than 20 instances in the United Kingdom where lawyers have used genetic data in support of the opinion that their clients are naturally hardwired for criminal behaviour.[16] In 2009, an Italian court issued a lighter sentence for a convicted murderer, partly because he had the 'bad' version of the *MAO-A* gene.[16] Two years later, another Italian court changed the sentence of a different convict (who had 'killed her sister, burned her corpse and attempted to kill her parents') from life to 20 years, based on *MAO-A* gene tests and neuroimaging.[17]

A glance over published research on *MAO-A*, however, demonstrates how ambiguous the situation around this gene is and highlights the scientifically, philosophically, and broadly socially problematic nature of the deterministic gene-centric outlook. There are data associating low levels of the active enzyme encoded by the *MAO-A* gene to certain cognitive and behavioural conditions, and there are data refuting that link.[18,19] Men with low MAO-A levels, for instance, were reported to make 'better financial decisions under risk', and low expression of *MAO-A* was significantly associated with greater happiness in women.[20–22] The expression of the 'warrior gene' and the properties of its product have been shown to be influenced by stress, diet, smoking, parental care and the social environment.[20,24] Provocation, history of child trauma, and systematic maltreatment were found to be very important for the onset of *MAO-A*-related aggressive, antisocial behaviour.

In 2012, molecular genetics again reached the news in connection to criminal behaviour. This time the focus fell on Adam Lanza, the mass murderer responsible for the Sandy Hook Elementary School shooting in Connecticut. Scientists, journalists and commentators expressed mixed feelings and worry over the announcement that the state's medical examiner had asked specialists at the University of Connecticut to analyse the gunman's DNA and search for genetic clues to his mindset. A quote from an editorial in the journal *Nature* illustrates the reaction to this development and critically addresses the reductionist and deterministic gene-centred view about criminality and human behaviour in general:

> But there is a dangerous tendency to oversimplify, especially in the wake of tragedy. If Lanza's DNA reveals genetic variants — as it inevitably will — people who carry similar variants could be stigmatized, even if those variants are associated only with ear shape. If Lanza has genetic variants already associated with autism or depression, people with those diseases could come under suspicion as well. The real risk here, and the real flaw in the Connecticut exercise, is that to identify a genetic variant is more straightforward — but arguably less informative — than to characterize the complex environment of the individual.[25]

The genome is just one of many factors influencing behaviour. Analyses have shown that about 50% of the variance in antisocial behaviours can be explained by genetics.[26] Nearly the same percentage has been calculated for the weight genetic factors have in human happiness, and lower figures in educational attainment.[22,27,28] Genome-wide association studies (GWAS) have not been very successful in finding solid links between genetic variants and complex phenotypes such as behaviour.[29] It is naturally difficult to delimit the contribution of the genetic aspect alone because the specificities of human behaviour are not necessarily determined by a single genetic unit, but rather by a whole constellation of genes, external and internal regulatory agents and other factors. The genome is just one strand in what Richard Lewontin calls 'The Triple Helix' of genes, organisms and environment.[30] To reflect this complexity, a multidimensional view of heredity has been proposed, which encompasses not only the genetic level but other biological, behavioural and cultural dimensions.[31]

Yet, although science and education professionals are familiar with this nuanced intricacy, they systematically tend to use simplistic, reductionist concepts and language, which, in turn, become inflated and broadly disseminated by the mass media. Evidently, genetic–deterministic outlooks, beliefs and explanatory frameworks remain widespread among scientists, teachers, students and the wider public.[32] One of the reasons for the allure and success of the bio-reductionist approach is its consonance with the deep-seated human yearning for simple answers to difficult existential questions. In the search for manageable essences and controllable primal causes, it is easy to see the orderly realm of molecular genetics as an enchanted source of such convenient answers and ready-to-apply strategies. The development of the science of heredity is inextricable from strong beliefs that the production of genetic knowledge will not only lead to profound understanding of the nature and essence of the living things, especially humans and human society, but the application of this knowledge will also allow their orderly and predictable management and improvement. This combination of applicability and the modernist urge to look for abstract, simplified, reductionist underpinnings and representations of a complex reality communicates constructively with the fact that 'hard' genetics predicts phenotypes generally better than a 'softer', more inclusive and multifactorial approach. Better predictability of outcomes translates into a higher, more privileged position in science, the social discourse, the public imagination and the technocratic dreams of controlled biological perfection.[14]

The interplay between scientific developments, psychological influences, ideological concerns and practical applicability characterising the formation of the determinist gene-centric outlook is unsurprising: genetics has always been intimately connected to plant and animal breeding, farming, medicine, anthropology and social engineering where technical–functional understanding, modelling and improvement of humans and other species are centre stage. The deterministic molecularisation of genetics successfully responds to the demands of this reality by reducing the complexity of heredity (and life at large) to relatively simple and manageable information-carrying-and-delivering fundamental causative entities. Thus, the simplified, reduced and abstracted in a modernist fashion deterministic gene becomes a convenient epistemic tool fruitfully employed in the study of biological inheritance, and in understanding and improving human society.

Interestingly, the very developments in molecular biology, biochemistry and biophysics associated with the surge and success of the reductionist–deterministic view of inheritance and the understanding of the gene as a manipulable physico-chemical program, are also central to the emergence of more complex and dynamic outlooks on heredity.

'Soft' Inheritance

Towards the end of the 20th century and the start of the new millennium, general advances in molecular biology and the sequencing of the human and other genomes began to disclose unforeseen genetic complexities. The arrangement of nucleotides making up the fertilised egg's DNA was not found to convincingly account for as much about the biology, behaviour and life of the individual and its progeny as anticipated. Just before the publication of the human genome sequence it was widely speculated that the genes which 'contain' the information 'to make a human being' would number 100,000, if not more. This figure, however, appeared to be greatly exaggerated. It transpired that the enormous complexity of *Homo sapiens* had to be associated with only around 20,000 genes — roughly as many as those located in the previously sequenced genome of the 1 mm long roundworm *Caenorhabditis elegans*. Furthermore, many of the growing number of examples of 'unusual' flows of molecular biological information had become relatively well-understood and integral to the genetic picture. Thus, at the end of 'the century of the gene', not only Crick's *Central Dogma of Molecular Biology*, but gene-centrism as a whole, needed to be re-examined.

Presently, the habitual understanding of the genotype–phenotype dyad as fixed within the strict lines of direct deterministic causation (where the genome is invested with some form of isolated information primacy and superiority), although still in wide circulation, has become increasingly inadequate in addressing the nature of heredity. A more open, multi-layered, probabilistic concept centred on interactive networks and systems has begun to look more appropriate. The focal range in the study of inheritance is now expanding to encompass not only the genes/genetic sequences and their direct, linear effects but also the intricacy of the cell, the organism, development, behaviour and experience. Investigations into the dynamic interaction between the hereditary system and the environment in the context of development and life experience are among the core drivers of this expansion.

Despite sustained ongoing efforts (*e.g.* sequencing and GWAS), the estimated heritability of complex phenotypes, including diseases and behaviours, has not been satisfactorily attributed to specific genetic variants. A great deal of heritability remains unaccounted for.[33] Moreover, changes in the DNA sequence alone have not been found to adequately justify the phenotypic response to rapidly changing environments and the swift formation of new heritable traits in one or several generations.[3] Some environmentally-induced plastic phenotypes have been shown to become genetically assimilated relatively quickly in the following generations and exhibited without the initial environmental stimuli. The reason for this plasticity and the overall 'missing heritability' could partly reside in the very multidimensionality of heredity where the DNA sequence is just one of many aspects.

Various observations and experiments demonstrated that relatively stable changes in gene activity, not linked to DNA sequence alterations, could occur during the lifetime of an individual, or more strikingly, even cross the limits of a generation and be passed on to the progeny.[29,34-36] It has become obvious that what is inherited from the parents is not just the genetic nucleotide 'script' but also a host of other layers of already present or acquired 'information'. A new scientific field studying the changes in the expression of the genetic program above (*epi-*) and beyond the DNA sequence emerged: epigenetics.

The term *epigenetics* was coined by the British polymath Conrad Hall Waddington in the 1940s to outline 'the branch of biology that studies the causal interaction between genes and their products, which bring the phenotype into being'.[37] The name is a combination of *epigenesis* (the Aristotelian understanding that development is a result of an unfolding sequence of interconnected steps, opposed to *preformation*) and *genetics*. Waddington's concept of epigenetics is positioned within a developmental context and attempts to bridge the dichotomy between genotype and phenotype drawn earlier by Johannsen. It does so through the incorporation of a connecting network of processes called *epigenotype* and the introduction of the image of a multidimensional *epigenetic landscape* which represents the relationship between the genetic aspect, the environmental influences and various possible developmental outcomes.

The coverage of the name 'epigenetics' has significantly mutated since its initial, development-oriented formulation and David L. Nanney's subsequent narrower delineation centred on gene expression, but it retains its original association with the strata beyond the genetic makeup, which interact with the genes and participate in the manifestation of the phenotype. Gradually, the concept of epigenetics became more molecularised. The epigenetic traits are currently seen as stably heritable phenotypes resulting from molecular changes other than alterations in the DNA sequence.[38] Such changes can be induced by internal (genetic, cellular and developmental) and/or external factors (food, stress, parenting, environment) and include DNA methylation, alterations of histone proteins and the effects of RNA molecules. The heritability of the produced changes in gene expression is mitotic, or sometimes, meiotic. It is precisely the transgenerational inheritance of environmentally-induced epigenetic effects that has attracted a great deal of scientific and popular interest in the recent years.[3,39]

Plentiful research data from a wide range of organisms have implicated epigenetics in cellular differentiation and specialisation, development, establishment of body architecture, ageing and fruit formation. Epigenetic phenomena are also involved in the stability and defence of the genome. The capacity to interact with the conditions of the macro- or micro-environment makes the epigenome an important element in the organism–environment interplay. Numerous links between environmental influence, epigenetic features (generational or transgenerational) and lasting functional or even structural alterations in the genome have been reported.[40–42] Horizontally, the epigenetic changes and their effects could be localised within a tissue (or tissues) of an individual organism or could be present in a group of individuals; vertically, these changes could be restricted to a generation or passed on to the progeny and the following generations.

Much of the understanding in epigenetics comes from research on plants and microorganisms but the last two decades have seen a dramatic increase in studies on animals and humans.[41,42] Particular attention has been drawn to the epigenetic influence of the socio-economic environment and various stress conditions (nutritional, psychological, many types of maltreatment, etc.), especially if the observed effects were transgenerational.[43,44] The long-term changes in the epigenomes of Holocaust or wartime famine survivors and their progeny have been widely discussed.[45,46] An expanding array of genes, among which is *MAO*, have been found to undergo maltreatment-, dietary-, and generally, lifestyle-dependant epigenetic modifications.[20,47]

Environmental transgenerational epigenetic inheritance, however, remains not well-understood, especially in vertebrates. Similarly, the process of genetic stabilisation of epigenetic modifications, and the epigenetic component of the 'assimilation' of environmentally-induced, plastic phenotypes have yet to be satisfactorily clarified.

Despite these uncharted areas, the discoveries in gene regulation and epigenetics have given a new dimension to the very meaning of the term 'hereditary information', extending it beyond the DNA blueprint, and particularly the narrow, simplistic confines of gene-centric determinism. Epigenetics and 'soft' inheritance confront fundamental notions in the field of genetics and biology, such as Johannsen's distinction between *genotype* and *phenotype*, Weismann's delimitation of *germplasm* and *soma*, and Francis Galton's high contrast separation of *nature* from *nurture*. Development, experience, behaviour, culture, and the influence of the environment, alongside a general flexible openness of the multiplex hereditary system, are becoming integral elements in this ever so complex mosaic.

Found to resonate with aspects of Lamarckism, such as the inheritance of acquired characteristics and the important shaping role of behaviour, the findings in epigenetics have prompted a revival of Lamarckian thought.[38] Even abandoned *vitalist* theories of life have made a comeback. This neo-Lamarckist upsurge is very noticeable in the social sciences and the humanities where it coincides with a renewed interest in the interaction between *the social* and *the biological*. Epigenetics has become a cross-disciplinary hub where different interests and ideologies creatively meet. That is why the developments in this field have to be understood within a broader social, intellectual and political scheme.

To an extent, the attractiveness, receptiveness and ideological useful-ness of epigenetics lie in the ambiguous nature of the term itself, which readily accommodates various convenient interpretations. A related, and perhaps deeper, reason for the success of this discipline is the counterpoint it is perceived to offer to the fatalistic understanding of genetics or biology as destiny. Epigenetics, it seems, promises liberation from the confines and necessities of genetic predetermination. In consonance with Waddington's expectations, the epigenetic sphere has become an emblem of anti-preformationism, a symbol of bio-environmental contextualisation, complex-ity, dynamic interaction and flexibility but not of atomisation, separation and rigidity.[48] As if the ossified hopelessness many saw in 'hard' heredity has now started to melt away exposing a new, plastic biological world of shifting impermanence, pliability, manifold opportunities and troubling vulnerability. Inflated optimistic messages centred on the promissory negation of genetic determinism and the associated view of biology as destiny can be easily located in the press. A main article in *Discover Magazine*, for instance, declared 'DNA Is Not Destiny: The New Science of Epigenetics. Discoveries in epigenetics are rewriting the rules of disease, heredity and identity'.[49] Some years later, *Time Magazine* responded with a similar cover story 'Why your DNA isn't your destiny' and *Der Spiegel*'s front page gloriously stated 'The Victory Over the Genes. Smarter, healthier, happier. How we can outwit our genome'.[50–52] Joining the chorus, *The Telegraph* thundered 'Epigenetics: How to alter your genes' while *Nature*'s 'The Sins of the Father' and *The Economist*'s 'Grandma's Curse' contributed darker overtones to a constantly growing list of dazzling titles.[53–55]

Epigenetics is often understood as indicating and embodying a break not only with genetic–biological tradition *per se*, but with tradition (something relatively stable which is transmitted, from the Latin *tradere*) in general. Interestingly, however, the rhetoric of biological plasticity, deconstruction and transformation, which celebrates the victory over genetic–biological destiny, coexists with and even incorporates simplistic deterministic messages. The dialectical concurrence of conflicting, yet woven in surprisingly similar terms, narratives of epigenetic flexibility and genetic rigidity ('gene for' one or another complex trait) is symptomatic of the dynamics in the field of genetics and the way heredity is currently understood.[3] A deeper look at this outward polarity and the underlying network of psychological, linguistic, and ideological connections could function as a starting point in a wider investigation of the relationship between genetics and epigenetics (symbolising 'hard' and 'soft' inheritance), which might contribute to a clearer and more synergistic understanding not only of this dyad but of the broader *nature/nurture* conflict.

In Search of Equilibrium

Modern epigenetics is largely defined by the understanding that the mitotically or meiotically heritable epigenetic differential gene expression is at heart something other than the sequence of nucleotides constituting DNA. An impression is generated that the epigenome is distinct, 'essentially' different from the genome. The epigenetic phenotypes, however, are linked to specific biochemical alterations of the DNA molecules or their immediate environment. Naturally, these alterations are not independent of the DNA nucleotide sequence and molecular constitution, or the structure of the involved histone proteins. Many epigenetic modifications are 'programmed', genetically determined, playing a particular spatio-temporal biological role. Moreover, the epigenetic marks influence DNA not only functionally but also structurally, potentially even in terms of nucleotide sequence, the DNA 'code' itself.[40] Thus, the position of the epigenome emerges as rather complex: dynamically interconnected with and situated between the molecular genome and the micro- or macro-environment as well as between the genotype and the phenotype.[3]

It has to be noted that this molecularised definition of the epigenome differs from the original, broader and less molecular-reductionist, Waddingtonian concept. Epigenetics has followed in the footsteps of genetics where the *gene* from a wider, vaguer, ontologically less fixed and perhaps more inclusive concept (*unit* of heredity), became more narrowly defined, and ontologised as a sequence of DNA nucleotides. Despite being contrasted as fundamentally different, the genome and the epigenome are, in fact, currently viewed in strikingly similar molecular and digital terms. In both cases, a wider complexity has been reduced to discrete, calculable and manageable essences, which can be located, isolated, examined and linked to predictable biological effects. This reductionism is of obvious epistemological value — both the deterministic molecular gene and the molecular epigenetic feature are useful conceptual tools in the study of heredity, development, and the interactions between the living systems and the environment. Yet, the notional restriction which such abstraction and utilitarian simplification engender cannot but carry various degrees of bias, so that as a whole these constrained and reduced concepts inevitably become insufficiently adequate in reflecting the reality they are supposed to represent. The formation of these molecularised tools (and the general development of both 'hard' and 'soft' inheritance) is ideologically laden and inseparable from the influence of expansive mechanistic visions of directed biological change leading to wider (social or other) effects.[14] In this light, the deterministic genome (DNA sequence) and the plastic epigenome share analogous conceptual contours: relatively simple, manageable molecular structures explaining heredity and life, which offer useful, predictable and controllable, transformative potential.

Lamarckism and 'soft' inheritance have a recognisable history of communication with progressivist theories aiming at controlled biological–social change and improvement.[39] Various 'softer' outlooks on heredity have always had a strong theoretical and practical appeal by suggesting that the living world is permeated with an immanent tendency or drive for perfection. This implication, combined with a focus on inherent direct environmentally- and behaviourally-responsive biological flexibility promises relatively quick, targeted and employable heritable changes. The notion of essential multifactorial flexibility of *nature*, however, comes at the expense of the predictability of the phenotypic outcome: the actualisation of the 'soft' transformative potential is difficult to productively control within a dynamic network of many elements. A significant driver of the *soma–germplasm* separation and the establishment of the reductionist–deterministic gene as a concept was the urge to challenge and invalidate Lamarckism and *vitalism* precisely for their innate obscurity. These theories were very difficult to experimentally prove, constructively use as epistemological instruments, and practically implement. Consequently, with the rise of population genetics and the advances in molecular biology, the inheritance of acquired characteristics and the 'soft' understandings of heredity became sidelined from mainstream science. The 'hard', less complex gene-centred views of genetics, which offered more clarity and relatively better predictability and control over the phenotype, gained overwhelming dominance.

Yet, this anti-Lamarckist and anti-*vitalist* shift did not lead to desertion or reassessment of the associated ideological outlooks and agendas of wide-reaching, controlled biological transformation and improvement. On the contrary, it has been shown that the pioneers of population genetics and the architects of the Modern Synthesis or neo-Darwinism (where Darwinism was combined with Mendelian-, Morganist-, and later, molecular genetics) were strongly motivated by ideas of targeted modification and perfection of humans and other species — the endurance of eugenics (in all its manifestations) being a striking example.[14,56] It is difficult to understand the establishment of 'hard' reductionist molecular determinism without considering the influence of such ideological factors and the necessity to convince and inspire the public, academia and industry to believe that the manipulation of a relatively simple molecular entity could have profound, predictable and manageable phenotypic effects.

Likewise, the modern-time reduction of the epigenome from a broad metaphor to a set of discrete, manageable biochemical features which modify the deterministic potential of the molecular genome (and are themselves determined by the genome itself and/or the environment) is not uncoupled from the influence of pervasive visions for biological–social change. As mentioned above, numerous media articles and academic works claim that we have finally been freed from the yoke of the genes and many fundamental problems, which have looked too complex to tackle, can now be easily understood, manipulated and solved through epigenetic tweaking of gene activity. New-school 'soft' genetics is expected to quickly penetrate, change and improve not only the state of the non-human or human individual or group of individuals but also this of humankind and living nature as a whole — curing disease, solving perennial problems, rectifying defects, strengthening weaknesses, optimising performance and creating wellbeing across the whole living spectrum — ominously resembling the way old-school 'hard' genetics was employed in the understanding, purification and medication of the individual genome or the collective (national, ethnic or racial) super-genome.[3]

From Galton's eugenic improvement of humanity, via the optimistic dreams of personal and social biological engineering shared by Julian Huxley, Linus Pauling, Joshua Lederberg, Francis Crick, James Watson, François Jacob, prominent researchers in the Human Genome Project, and many other key figures in the development of genetics as a science, to the current rhetoric of epigenetic betterment, the narrative of possible control and directed alteration through constructively reaching the very fabric and inner structure of the living world, especially of humanity, remains constant.[14] Such visions for engineered biological change leading to extensive physical, psychological, moral, and social transformations stem from a mechanistic understanding of nature and life and depend on convenient (and convincing) reductionist simplifications of vast, dynamic complexities. Thus, the current cycle of the oscillation between 'hard' and 'soft' views of heredity cannot be seen as exemplifying deep conceptual renewal, but rather as a continuation of the approach of mechanistic simplification where camouflaged molecular-reductionist metaphors and understandings, rooted in the tradition of genetic determinism, are transmitted and reused. 'At this [molecular] level', writes Nikolas Rose, 'anything and everything appears, in principle, to be intelligible, and hence to be open to calculated interventions in the service of our desires about the kinds of people we want ourselves and our children to be'.[53] To fulfil this creative biomolecular potential and ambition, the 'new' epigenome like the 'old' genome has to be imagined, understood and ontologised reductionistically–deterministically: if we decipher and know all items constituting the structure and nature of the epigenetic dimension of heredity, we can predict the associated outcomes and effects, or in the spirit of Laplace's thought, we can have the future and the past of the living things present to our eyes. This knowledge can then be instrumentalised and applied as we will.

Such useful simplistically-deterministic explanations are manifest features of the rhetorical epigenetic landscape, where the promise for flexibility and transformation coexists with understandings of fixed 'code', 'instructions' and 'program'.[3,57] The structural/biochemical interconnectedness of genetics and epigenetics is inseparable from the linguistic/semantic one. This relationship can be 'apophatic' (negativising genetics: epigenetics is what 'hard' inheritance is not) or non-negatory (borrowing directly from the deterministic linguistic arsenal of genetics, such as language of information, control and simplistic molecular causality). The epigenetic modifications are presented and understood as relatively-easy-to-grasp molecular changes which *control* the expression of the genetic *program*. They *determine* the actualisation of the *deterministic* potential of the genes (DNA sequences) and are themselves *controlled* and *determined* environmentally and/or genetically. Epigenetics has become an integral element in the molecular and rhetorical circle of biological determinism.

Forming convenient theoretical and practical biological tools, such as the molecularised genome or epigenome, is not only essential in mechanistically studying and understanding the living world but it also enables the human will to successfully mould and improve non-human and human life — augmenting, purifying and protecting the desired aspects and even constructing new useful attributes. The creative principle is fundamental to the way biological heredity is approached and dealt with. It precedes the differentiation of genetics as a science and is central to millennia of plant and animal domestication, breeding and farming. As the genetic abstraction became theoretically framed and clothed in molecular structure, the science of heredity obtained a pivotal epistemological and practical position in the creative process of comprehensive biological and social optimisation. Reflecting the general *nature/nurture* dichotomy, the specific understandings and strategies for the process of alteration and improvement, however, began to significantly differ. 'Hard' geneticists thought that the stable, transmittable elements forming the genotype can be selectively rearranged or even created anew. The proponents of 'soft' inheritance saw heredity as a complex, less stable and more dynamic system strongly influenced (determined) by the changing environmental factors.

Thus, although diverging in key details but united in their strong connection to modification and fashioning of the world led by creative human will, intellect, and practical knowledge, both genetics and epigenetics can be interpreted partly through the metaphor of *Homo faber*.[58] This concept, which views humans as creative agents controlling their environment and fate, was salient in Renaissance humanism but is also articulated in the work of Henri Bergson, Max Scheler, Max Frisch, Hannah Arendt and others. The inherent creative human impulse surges, and stimulating observation, studying and comprehension of nature, finds its conclusive realisation in influencing the living world. This it achieves by creatively manipulating heredity. Such manipulation, however, depends on deterministic and reductionist mechanistic approaches as it has to be functional, reliable and predictable. Mimicking the Divine Creator, *Homo faber*, the creator of tools to control and construct his or her surroundings, body, mind and destiny, imagines, discovers and forges the simplistic, reductionist–deterministic underpinnings of heredity to facilitate, focus and actualise the creative urge and process which he/she carries and enacts in pursuit of the ultimate goal: management and improvement of life. This is particularly clear in the thought-frame of 'hard' genetics or in the conceptual, rhetorical and symbolic dimension of the molecularised scientific study of heredity in general (*e.g.* viewing genes as life 'essences' and organisms as 'survival machines' for the genes; deciphering and altering the 'book of life' which contains the 'gene for' every feature, the genetic 'blueprint', 'program', 'instructions', and 'destiny' through calculated 'selection', 'eugenics' or 'genetic engineering'.) The epigenetic (or 'soft') aspect of inheritance, although more multifactorial, probabilistic and deterministically flexible, also resides on this pragmatic 'creative' conceptual infrastructure. Here the environment (understood as a network of causative factors and effects) influences in an intelligible and predictable chemical fashion the hereditary foundation through a dynamically-interconnected with the genome and the environmental conditions *epi*-genetic stratum perceived in convenient, simplistic, molecular terms. So to be explained and harnessed, the deterministic effects of the environment are again reduced to manageable biochemical relationships, processes and structures.

The epigenetic sphere carries a striking cyclical paradox: the execution of *free creative will* leads to epigenetic modifications, which can influence and restrict the behaviour and *free creative will* not only of the receiving individual or group of individuals but also of the acting individual or group. Development, behaviour, choice and the social environment are all now understood as biochemically–epigenetically determining various phenotypes but also as themselves being epigenetically determined. Many complex characteristics are increasingly becoming epigenetically biologised.[59] The great emphasis placed on social, racial and gender characteristics draws deterministic angles with profound ethical and legal consequences.[60]

Evidently, biological determinism mutates but remains strong in the wider social sphere and academia: genetic deterministic beliefs are being exchanged or supplemented with molecularised environmental or behavioural deterministic views. Various authors discern a common spirit of epigenetic determinism permeating the current epigenetic and even wider biological and social discourse.[57] The deterministic dimension of epigenetics is apparent in the heightened attention and the privileged position which the transgenerational epigenetic effects receive in this discourse. Although the documented examples of environmental transgenerational epigenetics, especially in mammals, are relatively few (the vast majority of epigenetic phenotypes being relatively transient), the more permanent and persistent in the progeny the epigenetic modifications are found to be, particularly in humans and other mammals, the more intriguing scientifically and socially they are. Consequently, if the epigenome is demonstrated to 'internalise' environmental influences into stably heritable differential gene expression associated with important (desired or unwanted) phenotypic effects, it will automatically obtain a special biological–social status and will have to be revered and protected like the genome. This lust for protection, purity and improvement, embodied graphically in eugenics, could lead to the development of equally problematic *epieugenics* or *euepigenics*.[3,60]

The interwoven but opposing narratives of plasticity and stability characterising the epigenetic discourse resonate with different understandings of nature and existence: the former narrative communicates with broadly existentialist and postmodernist concepts of fluidity, while the latter narrative — with classical, *traditional* ideas where a relatively stable, universal aspect exists (*essence, nature*) and is transmitted (as in *tradere*). Seeing epigenetics as centred on plasticity and environmental modifiability, for instance, is naturally open to communication with the trope of *existence-over-essence*. Through the epigenetic angle, identity (or perhaps *nature*) can look biologically very malleable and fluid, paralleling the idea that *nurture*, experience and the external factors, determine *essence* and consciousness.[61] If something *essential* is present and transmitted at all (can modifications occur and persist without a basis which is modified and transmitted?), it is mouldable and ever-shifting. The understanding and the presentation of environmental epigenetics in some measure echo the existentialist–postmodernist scepticism and deconstructive attitude towards fixed inner meaning, universality and tradition. Perceived as permeable and flexible, the biological blends with the social, cultural, contextual and experiential. To a degree, such syncretism can be helpful and healing in terms of the wounds opened with the separation of *the biological* from *the social*, the radicalisation of the *nature/nurture* debate and the gene-centric outlook of heredity as destiny. Yet, the potential of this emerging symphony is not maturely harnessed but much of it is channelled to serve ideological and mercantile agendas involving comfortably simplistic, exaggerated and inadequate interpretations. The current situation is reminiscent of the radical move towards ideologically-influenced 'soft' inheritance in Soviet science associated with the work and anti-genetic rhetoric of Stalin's infamous chief agronomist Trofim Lysenko.[62] Although not identical with Soviet Lysenkoism, modern-day neo-Lamarckism shares similarities with the spirit of this doctrine, which are apparent in the hyperbolised depiction of environmental epigenetics as a magic wand for directed and relatively quick universal biological and social transformation.[39] In its current molecular-reductionist form, however, epigenetics does not only challenge bio-deterministic thinking but also encourages and includes such perceptions of life, experience and the human situation. Notions of 'code', 'program', 'instructions' and manageable causative molecular interactions with wide-reaching effects are prevalent in the epigenetic discourse while the 'internalisation', 'biologisation' or 'stabilisation' of external influences and the transgenerational 'permanence' of some epigenetic phenomena are vividly emphasised. Thus, swept in the confusion of ricocheting rhetorics of flexibility and stability, charged with social–constructivist stereotypes or essentialist–mechanistic over-biologisation, epigenetics is burgeoning, markedly influencing the debate and understanding of heredity and living nature while still maturing.

Despite the problems accompanying its growth, epigenetics could be helpful in moderating the current bipolar perspective of 'hard' and 'soft' inheritance through contributing a richer and more balanced outlook: a middle path, less susceptible to problematic, limiting determinism and reductionism, and closer to the classical understanding of the *nature/ nurture* duality. Potential steps towards this end could include: the return to a broader understating of the gene and its re-conceptualisation from a sequence of nucleotides to a *unit* of heredity (multilayered, including the genetic and epigenetic aspects); careful choice of language which does not reflect and prompt simplistic deterministic notions but relates to the multifactorial nature of inheritance; and a general portrayal of *the biological* as not radically separated from the environment and *the social*. To be more adequately understood, heredity has to be addressed along compatibilist lines of moderate determinism and indeterminism, mechanism and holism. Naturally, such a reconciliatory union cannot be without tension. From the views of the Ancients, through expanses of theological debates, to the existential dilemmas, challenges and limitations faced by Bergson's, Scheler's, Frisch's or Arendt's *Homo faber*, the questions about probability and determinism, free will and fate, mechanism and organicism remain salient and difficult to tackle, demanding a multifaceted, ecumenical approach where answers are not sought in opposing binomial terms but in synergism.

Ethical and Concluding Remarks

Reflecting the extensive *nature/nurture* dichotomy, the current academic and public heredity-related discourse is noticeably bipolar, characterised by opposing narratives of genetic stability and epigenetic flexibility. This apparent confrontation is rooted in tendentious interpretations of the biological and social reality where enduring common deterministic sentiments, understandings and representations are routinely recycled to fit overarching ideological pictures and agendas. Thus, albeit outwardly contrasted, the modern views of 'hard' and 'soft' inheritance share similar reductionist molecular–deterministic conceptual and rhetorical underpinnings. The developments in epigenetics, however, are not to be readily overlooked because of these interconnections or neglected in the general confusion surrounding their extravagant publicity. Being naturally open to multidisciplinary approaches, epigenetics possesses an appreciable potential to constructively address the conceptual and semantic limitations of the simplistic–deterministic outlooks on inheritance through an ecumenical approach reflexive of the complexity of heredity and thus contribute to the mitigation of the *nature/nurture* polarity and the antagonism between *the biological* and *the social*.

An important aspect of multidisciplinary interest is the wide biological and social attention which epigenetics draws to personal and collective responsibility. Although molecular–deterministic in their conception and presentation, the developments in epigenetics dialectically emphasise the role of free will, indicating that choice and conduct impact not only the current lives of humans and non-humans but also the existence and wellbeing of the future generations.[63,64] Behaviour and choice themselves are now viewed as both subjects and objects of epigenetic influence. Naturally, such powerful ethical aspects carry complex arrays of problems and susceptibilities to abuse, which require timely considerations by policy makers, researchers, educators, health specialists and legislators. The transformative epigenetic potential and the plastic understanding of heredity pose important questions concerning the reach and permanence of the epigenetic modifications and the extent of the 'softness' of inheritance. Relating strongly to taxonomy and systematics but also to identity, ethnicity, nation, race, character and individuality, these problems call for serious collaborative attention. The emerging complex and dynamic picture of heredity resounds with an emotive passage from Bergson's final work. Studying the sources of morality and religion, he writes: 'Let us then give to the word biology the very wide meaning it should have, and will perhaps have one day, and let us say in conclusion that all morality, be it pressure or aspiration, is in essence biological'.[65] Isn't it time for biology to become what it is supposed to be — the holistic study of life?

References

1. Laplace PS. (1902) *A Philosophical Essay on Probabilities*. J. Wiley and Sons, New York; Chapman & Hall, London.
2. Deichmann U. (2016) Epigenetics: The origins and evolution of a fashionable topic. *Dev Biol* **416**(1): 249–254.
3. Gadjev I. (2017) Epigenetics, representation, and society. *Zygon* **52**(2): 491–515.
4. Rheinberger HJ, Müller-Wille S and Meunier R. (2015) *Gene*. The Stanford Encyclopedia of Philosophy. Stanford University, CA.
5. Roll-Hansen N. (2011) *Lamarckism and Lysenkoism Revisited. Transformations of Lamarckism: From Subtle Fluids to Molecular Biology*, 1st ed. pp. 77–88.MIT Press, Cambridge, Massachusetts.
6. Loison L. (2015) Why did Jacques Monod make the choice of mechanistic determinism? *C R Biol* **338**(6): 391–397.
7. BBC. (2012) Binge-drinking gene discovered. *BBC News*. http://www.bbc.co.uk/news/health-20583113 [18 April 2018].
8. Coghlan A. (2011) Teen survey reveals gene for happiness. *New Scientist*. http://www.newscientist.com/article/dn20451-teen-survey-reveals-gene-for-happiness.html [18 April 2018].

9. Connor S. (1995) The 'Gay Gene' is back on the scene. *The Independent*. http://www.independent.co.uk/news/the-gay-gene-is-back-on-the-scene-1536770.html [18 April 2018].

10. Ellwood M. (2013) The genetics of bargain hunting. *Time*. http://ideas.time.com/2013/10/21/the-genetics-of-bargain-hunting/ [18 April 2018].

11. Langone J. (2004) In search of the 'God Gene.' *The New York Times*. http://www.nytimes.com/2004/11/02/health/02book.html?_r=2& [18 April 2018].

12. MacArthur DG and North KN. (2004) A gene for speed? The evolution and function of Alpha-actinin-3. *BioEssays* **26**(7): 786–795.

13. Carter C. (2015) Women are more likely to cheat on their partner if they carry the 'infidelity gene', scientists discover. *Daily Mail*. http://www.dailymail.co.uk/sciencetech/article-2954349/Women-likely-cheat-partner-carry-infidelity-gene-scientists-discover.html [18 April 2018].

14. Esposito M. (2017) Expectation and futurity: The remarkable success of genetic determinism. *Stud Hist Philos Sci C* **62**: 1–9.

15. Brunner H, Nelen M, Breakefield X, *et al.* (1993) Abnormal behavior associated with a point mutation in the structural gene for monoamine oxidase A. *Science* **262**(5133): 578–578.

16. Feresin E. (2009) Lighter sentence for murderer with 'Bad Genes.' *Nature*. https://www.nature.com/news/2009/091030/full/news.2009.1050.html [18 April 2018].

17. Owens B. (2011) Italian court reduces murder sentence based on neuroimaging data. *Nature*. http://blogs.nature.com/news/2011/09/italian_court_reduces_murder_s.html.

18. Kunugi H, Ishida S, Kato T, *et al.* (1999) A functional polymorphism in the promoter region of monoamine oxidase — A gene and mood disorders. *Mol Psychiatr* **4**(4): 393–395.

19. Ficks CA and Waldman ID. (2014) Candidate genes for aggression and anti-social behavior: A meta-analysis of association studies of the 5HTTLPR and MAOA-uVNTR. *Behav Genet* **44**(5): 427–444.

20. Bortolato M and Shih JC. (2011) Behavioral outcomes of monoamine oxidase deficiency: Preclinical and clinical evidence. *Int Rev Neurobiol* **100**: 13–42.

21. Frydman C, Camerer C, Bossaerts P and Rangel A. (2011) MAOA-L carriers are better at making optimal financial decisions under risk. *P Roy Soc B-Biol Sci* **278**(1714): 2053–2059.

22. Chen H, Pine DS, Ernst M, *et al.* (2013) The MAOA gene predicts happiness in women. *Prog Neuro-Psychoph* **40**: 122–125.

23. McDermott R, Tingley D, Cowden J, *et al.* (2009) Monoamine oxidase A gene (MAOA) predicts behavioral aggression following provocation. *Proc Natl Acad Sci USA* **106**(7): 2118–2123.

24. Lea R and Chambers G. (2007) Monoamine oxidase, addiction, and the 'warrior' gene hypothesis. *New Zeal Med J* **120**(1250): U2441.

25. Editorial. (2013) No Easy Answer. *Nature* **493**(7431):133.

26. Beaver KM, Wright JP, Boutwell BB, *et al.* (2013) Exploring the association between the 2-repeat allele of the MAOA gene promoter polymorphism and psychopathic personality traits, arrests, incarceration, and lifetime antisocial behavior. *Pers Indiv Differ* **54**(2): 164–168.

27. Rietveld CA, Cesarini D, Benjamin DJ, *et al.* (2013) Molecular genetics and subjective well-being. *Proc Natl Acad Sci USA* **110**(24): 9692–9697.

28. Lee JJ and consortium. (2018) Gene discovery and polygenic prediction from a genome-wide association study of educational attainment in 11 million individuals. *Nat Genet* **50**(8): 1112–1121.

29. Daxinger L and Whitelaw E. (2012) Understanding transgenerational epigenetic inheritance via the gametes in mammals. *Nat Rev Genet* **13**(3): 153–162.

30. Lewontin RC. (2000) *The Triple Helix: Gene, Organism, and Environment*, 1st edn, Harvard University Press Cambridge, Massachusetts.

31. Jablonka E and Lamb MJ. (2014) *Evolution in Four Dimensions: Genetic, Epigenetic, Behavioral, and Symbolic Variation in the History of Life (Revised Edition)*, 2nd edn, MIT Press, Cambridge, Massachusetts.

32. Carver RB, Castéra J, Gericke N, *et al.* (2017) Young adults' belief in genetic determinism, and knowledge and attitudes towards modern genetics and genomics: The PUGGS questionnaire. *Plos One* **12**(1): e0169808.

33. Maher B. (2008). The case of the missing heritability. *Nature* **456**: 18–21.

34. Crews D, Gillette R, Miller-Crews I, *et al.* (2014) Nature, nurture and epigenetics. *Mol Cell Endocrinol* **398**(1–2): 42–52.

35. Heard E and Martienssen R. (2014) Transgenerational epigenetic inheritance: Myths and mechanisms. *Cell* **157**(1): 95–109.

36 Tollefsbol T. (ed.) (2014) *Transgenerational Epigenetics*. Academic Press, London, Waltham, San Diego.

37. Jablonka E and Lamb MJ. (2002) The changing concept of epigenetics. *Ann NY Acad Sci* **981**: 82–96.

38. Berger SL, Kouzarides T, Shiekhattar R and Shilatifard A. (2009) An operational definition of epigenetics. *Gene Dev* **23**(7): 781–783.

39. Gadjev I. (2015) Nature and nurture: Lamarck's legacy. *Biol J Linn Soc* **114**(1): 242–247.

40. Boyko A and Kovalchuk I. (2011) Genome instability and epigenetic modification-heritable responses to environmental stress? *Curr Opin Plant Biol* **14**(3): 260–266.

41. Skinner MK. (2014) Environmental stress and epigenetic transgenerational inheritance. *BMC Med* **12**(1): 153

42. Schaefer S and Nadeau JH. (2015) The genetics of epigenetic inheritance: Modes, molecules, and mechanisms. *Q Rev Biol* **90**(4): 381–415.

43. McGuinness D, McGlynn LM, Johnson PCD, *et al.* (2012) Socio-economic status is associated with epigenetic differences in the pSoBid cohort. *Int J Epidemiol* **41**(1): 151–160.

44 Ramo-Fernández L, Schneider A, Wilker S and Kolassa IT. (2015) Epigenetic alterations associated with war trauma and childhood maltreatment. *Behav Sci Law* **33**(5): 701–721.

45 Heijmans BT, Tobi EW, Stein AD, *et al.* (2008) Persistent epigenetic differences associated with prenatal exposure to famine in humans. *Proc Natl Acad Sci USA* **105**(44): 17046–17049.

46 Yehuda R, Daskalakis NP, Bierer LM, *et al.* (2015) Holocaust exposure induced intergenerational effects on FKBP5 methylation. *Biol Psychiat* **80**(5): 372–380.

47. Melas PA, Wei Y, Wong CCY, *et al.* (2013) Genetic and epigenetic associations of MAOA and NR3C1 with depression and childhood adversities. *Int J Neuropsychop* **16**(7): 1513–1528.

48. Stotz K and Griffiths P. (2016) Epigenetics: Ambiguities and implications. *Hist Phil Life Sci* **38**(4): 1–20.

49. Watters E. (2006) DNA is not destiny: The new science of epigenetics. *Discover Magazine.* http://discovermagazine.com/2006/nov/cover [18 April 2018].

50. Der Spiegel. (2010) Der sieg uber die gene. *Der Spiegel.* http://www.spiegel.de/spiegel/print/d-73109479.html [18 April 2018].

51. Maher B. (2008) The case of the missing heritability. *Nature* **456**: 18–21.

52 Cloud J. (2010) Why your DNA isn't your destiny. *Time.* http://content.time.com/time/covers/0,16641,20100118,00.html [18 April 2018].

53. The Economist. (2012) Grandma's curse. *The Economist.* http://www.economist.com/news/science-and-technology/21565573-some-effects-smoking-may-be-passed-grandmother [18 April 2018].

54. Bell C. (2013) Epigenetics: How to alter your genes. *The Telegraph.* http://www.telegraph.co.uk/news/science/10369861/Epigenetics-How-to-alter-your-genes.html [18 April 2018].

55. Hughes V. (2014) The sins of the father. *Nature* **507**: 22–24.

56. Rose N. (2007) *The Politics of Life Itself: Biomedicine, Power, and Subjectivity in the Twenty-First Century,* 1st edn, Princeton University Press, Princeton, USA.

57. Waggoner MR and Uller T. (2015) Epigenetic determinism in science and society. *New Genet Soc* **34**(2): 177–195.

58. Lee K. (2005) Philosophy and revolutions in genetics: Deep science and deep technology, 1st edn., Palgrave Macmillan, Basingstoke, UK.

59. Gunter TD and Felthous AR. (2015) Epigenetics and the law: Introduction to this issue. *Behav Sci Law* **33**(5): 595–597.

60. Meloni M and Testa G. (2014) Scrutinizing the epigenetics revolution. *BioSocieties* **9**(4): 431–456.

61. Juengst ET, Fishman JR, McGowan ML and Settersten R. (2014) Serving epigenetics before its time. *Trends Genet* **30**(10): 427–429.

62. Maderspacher F. (2010) Lysenko rising. *Curr Biol* **20**(19): R835–R837.

63. Lock M. (2013) The lure of the epigenome. *Lancet* **381**(9881): 1896–1897.

64. Pickersgill M, Niewöhner J, Müller R, *et al.* (2013) Mapping the new molecular landscape: Social dimensions of epigenetics. *New Genet and Soc* **32**(4): 429–447.

65. Bergson H. (1935) *The Two Sources of Morality and Religion.* Doubleday Anchor Books/Henry Holt & Co., Garden City, NY, USA.

5 Organisms, Development, and Evolution: Invitation to a New Understanding

David J. Depew*

Introduction: The Vexed Situation of Evolutionary Biology

Modern science is a practice of inquiry in which quantitative methods are applied to, and held accountable by, empirical data: data acquired by experimentation or controlled observation conducted in the spirit of experimentation.

By this standard, physics is the premier science. By the 20th century, its many triumphs included bringing chemistry under its sway. Linus Pauling did just that when he deduced the laws of chemical bonding from quantum mechanics.[1] The young James Watson nursed a burning ambition to do Pauling one better by reducing biological heritability to biochemistry. How he and Francis Crick succeeded, albeit with involuntary help from Rosalind Franklin, is well known to most people.[2] In one fell swoop they discovered in the molecular structure of DNA the mechanism by which genes are passed to descendants and the basic source of the variation that provides natural selection with evolution's fuel: spontaneous random mutation in the base

*Emeritus Professor of Communications Studies and POROI, The University of Iowa, Iowa City, Iowa, USA.

pairs that hold DNA together.[3] The discovery brought the genetic theory of natural selection into close connection with more fundamental sciences, renewing hopes that eventually all fields of inquiry, including behavioural psychology, will be more systematically connected to physical principles.[4]

Why is unifying the sciences in this reductionist spirit so prized? In the remote background is the medieval theology of William of Ockham, who inferred from God's attribute of simplicity that he could be counted on to create a world with as few operating principles as possible. Ockham's razor, as it is called, was both honoured and subverted by efforts in 19th century Europe to promote science as a purely secular enterprise. Presenting the unity of science as proof of its ability to converge on objective truth was meant to suggest that scientists should be assigned a key role in policymaking in states free of clerical influence. The thought was that when power rests on knowledge rather than faith, Bacon and Descartes's dream of improving man's lot by techno-scientific invention would at last come true.

After World War II, History of Science programs and departments were founded in the expectation that primary sources would show that the 'logical empiricist' picture sketched above had been working itself out even before people were devoted to achieving it. But this isn't what happened. Historical reconstruction showed that the mathematical and empirical components of even the most well confirmed theories are glued together by a third factor, conceptual frameworks, which announce their presence in the images, similes, metaphors, and analogies that scientists use to interpret what they are doing not only to the public, whose consent and financial support they must court, but also to themselves. Philosophers of science were thus presented with a problem that still occupies them. Thomas Kuhn's notion of paradigms, which emerged from his work in the history of astronomy and physics, is the most well-known three-legged, rather than two-legged, picture of the structure of scientific theories. Like the makers of other meta-scientific models that followed, Kuhn realised that the mediating role of conceptual frameworks imposes a higher than expected burden of proof on scientific theories for the simple reason that commitment to a theory precedes the data that can prove it. Lifted on the wings of metaphor, pre-commitment readily inflates into worldviews that make it difficult even to see, let alone account for, non-confirming facts or anomalies.[5] It is impossible to view the epistemic credentials of evolutionary inquiry without taking into account the role of conceptual frameworks, metaphors, and runaway worldviews.

Species transformism existed before Darwin's seminal 1859 book gave it a plausible mechanism and some empirical support. From the start, however, Darwin's attack on natural theology — the intelligent design creationism of his day — entangled his theory with attempts to defend a secular worldview in a country in which knowledge was still under official religious influence. There is much continuity between the anti-clericalism of Darwinism's early defender T. H. Huxley and the anti-theism of Richard Dawkins.[6] Secularised biology has tended to encourage metaphysical materialism, ontological atomism (in which organisms are seen as assemblies of parts), theoretical reductionism, and an abiding hope that we can use genetic engineering to cure 'inborn errors of metabolism' and devise bio-enhancements that will put humans, or at least some of them, in the driver's seat of evolutionary advance, an ambition often figured as humans taking God's place.

Anyone who teaches evolutionary biology or is engaged in efforts to promote its public understanding will have directly experienced how questions about theology and biotechnology still get in the way of their efforts to explain the significance of recent research in the light of evolutionary dynamics. How this pedagogical difficulty makes itself felt today, and what can be done about it, is the motivating question of this chapter.

It may be comforting to recall that for a long time after it took off in the 17th century physics, too, was entangled with theology, metaphysics, and technological utopianism. Still, by the time of James Clerk Maxwell and Ludwig Boltzmann people could tell the difference between technical disputes among physicists and subjective opinions freely offered by the field's leading figures about everything from what God thinks to the putative benefits of socialism or eugenics. Today it is either false or hopelessly ambiguous to say that modern physics has justified the 'Newtonian worldview'. Does that mean that it defeated Cartesianism? The issue seems irrelevant now, or even forgotten. So why do evolutionists keep crowing that a 'Darwinian worldview' is about to triumph with the emergence of evolutionary psychology? Isn't it more likely that in the future, evolutionary science will honour Darwin in the same way physicists honour Newton, by burying him with abundant praise?

Evolutionary biology is a young science. So, perhaps all frustrated teachers and public lecturers have to do is wait patiently until students and citizens have grown accustomed to drawing distinctions as nuanced as those we routinely make between what physics has learned and what physicists opine. Unfortunately, however, we can't anticipate the future as if it were already or almost here. We must respond to the situation of evolutionary discourse as it presents itself to us here and now. Evolutionary inquiry has long since become a science — it systematically acquires and banks objective knowledge — but how to guess what might still be discovered and reinterpret what we already know continues to be affected by conceptual assumptions and ideologies.

Writers like Dawkins, E. O. Wilson, Daniel Dennett, and Steven Pinker still proclaim that the problem lies in the persistence of religion. For them the solution is resolute reductionism in science and equally resolute materialism. Their writings have a 19th century feel. Their reading of the history of evolutionary biology is not mine. Instead, I will argue that differences between the objects of physics, biology, and the human sciences have always limited the project of encoding evolutionary biology's discoveries in reductionist terms, and still do. Organisms and the ecological processes in which they are embedded, including social systems, are inherently many-layered and dynamically intertwined. What evolutionary biologists have learned about them has come from finding mathematical, experimental, and conceptual ways to make this complexity tractable. Read rightly, the record shows that taming biological complexity means learning to acknowledge, appreciate, and understand what organisms uniquely are, not seeing them as something else. This learning process is especially intense at present.

As a result, I believe that reductionistic rhetorics of biology are encoding increasingly less empirical knowledge and increasingly more conceptual and ideological posturing. Presenting evolutionary biology in the light of this history may help narrow the gap between public and professional understanding of evolution and place arguments about religion and bio-technology in a more judicious light.

Taming Complexity: Evolution After Darwin

In the 19th century, experimental embryology was the most advanced of the sciences of life. Many embryologists kept their distance from evolutionary speculation altogether, but those who took the leap favoured the idea that phylogeny is ontogeny writ large. New kinds, they hypothesised, evolve when a step in the developmental process is deleted or a new one is added.[7] Some advocates of ontogeny–phylogeny parallelism called themselves Darwinians. That was primarily because they took themselves to be defending Darwin's thesis of descent from a common ancestor. This fact, Ernst Haeckel argued, can be understood only if the universe is closed to intervention from the great beyond.

Virtually all the immediate forebears of 20th century genetic Darwinism were embryologists whose efforts to confirm ontogeny–phylogeny parallelism led to discrediting it. (This is not the last time we will observe this ironic pattern; it is more the rule than the exception in the history of evolutionary biology.) I have in mind people like Raphael Frank Weldon, who used statistical methods to confirm the existence of natural selection in the wild by correlating morphological changes in crabs with increased industrial pollution in Plymouth Bay; Hugo de Vries and William Bateson, who attempted to find in sudden mutations in Mendel's recently rediscovered discrete units of heredity what Weldon ascribed to gradual natural selection at work on continuous observable traits (phenotypes); and T. H. Morgan, who located these genes (as Wilhelm Johannsen called them) at specific points (loci) on physical chromosomes. These efforts reached an acme of sorts when R. A. Fisher — a statistical whizz with no knowledge of embryology — brought together genetic mutationism and Weldon's statistical approach to natural selection. He did so by demonstrating that the probability that a genetic macro-mutation will take hold in a population is vanishingly small, while a stream of heritable mutations each with a miniscule effect will do the job under the additive, amplifying, adaptive influence of natural selection.[8,9] By tracking genotypes rather than phenotypes, Fisher's probabilistic inference confirmed Weldon's conviction that 'The problem of ... evolution is essentially a statistical problem'[10] (p. 329). He also validated Darwin's idea that natural selection must work gradually if it is to have evolutionary consequences.

Fisher's 'fundamental principle of natural selection', together with several other ways in which gene frequencies can shift in populations — in particular, genetic drift, or chance fixation of genes in small populations, and gene flow from one population to another by migration — constitutes the probabilistic calculus at the heart of the Modern Evolutionary Synthesis. Erected in the 1940s and 1950s with the aim of unifying genetics with systematics, palaeontology, and other aspects of natural history, the Modern Synthesis has organised much of what we know about evolution. The Synthesis treats evolution as change in gene frequencies in Mendelian populations: populations open to genetic exchange at the reproductive instants that tie one generation to another.[11] Organisms develop, the Synthesis says, but don't evolve; populations evolve but don't develop. The embryologist Scott Gilbert notes that the phrase 'Modern Evolutionary Synthesis' implies an unspoken contrast with a prior 'unmodern' synthesis: ontogeny–phylogeny parallelism.[12] His point is that with the rise of the Modern Synthesis developmental biology, which didn't have a statistical bone in its body, lost its relevance to evolutionary inquiry — until recently.[13] The benefits of understanding the dynamic relationship between environment, organism, and gene from the level of statistically and probabilistically describable populations outweighed any direct connection between development and evolution. The shift was not unlike the way statistical mechanics and thermodynamics eclipsed classical deterministic physics.

Led by Ernst Mayr, the makers of the Modern Synthesis downplayed differences in their ranks in order to promote population-based evolutionary biology's status as a fully mature, objective, suitably (but not exclusively) quantitative 20th century science.[14,15] Attempts by analytic philosophers of biology in the 1960s formally to axiomatise the principles of population genetics further unified the genetic theory of natural selection and insulated it from the enthusiasm many of its progenitors had earlier shown for eugenics, sometimes (but not always) laced with traces of racism.[16,a] Emphasising unity disguised differences not only between individual scientists, but more importantly between two long-lived research projects to which the Synthesis quickly gave rise.

One of these projects, 'population ecology', was devoted to proving that traits previously used as species markers because they were assumed to be adaptively neutral (and hence stable enough to construct typological classifications) are actually adaptations that reflect the exigencies of particular environmental circumstances. By contrast, what I will call the 'speciation project' used the reform of systematics it shared with population ecologists to explore how Mendelian populations evolve genetic barriers to reproducing, how new species radiate biogeographically, and how higher *taxa* evolve. Richard Lewontin remarked that the seminal population genetic speciationist Theodosius Dobzhansky, under whom he trained, 'wouldn't have dreamed of spending a large amount of time studying one particular behavioural or physiological mechanism in one particular organism the way [adaptationists] did'[17] (p. 32). That is partly because he

---

[a]In the postwar period many authoritative geneticists believed that repudiating racism made scientific eugenics possible for the first time. Among them were Julian Huxley and Hermann Muller. Dobzhansky insisted that in population thinking eugenics and racism rise and fall together. He was contemptuous of Huxley and vigorously disputed Muller[16] (Dobzhansky to L. C. Dunn, April 26, 1947 — all correspondence cited is archived in the Theodosius Dobzhansky Papers, American Philosophical Society, Philadelphia, PA).

had bigger fish to fry. But it was also because in explaining the dynamics of large-scale evolutionary change, speciationists invoked mutation rate, genetic drift, and gene flow as well as natural selection, and as a result placed a much higher burden of proof on identifying a trait as an adaptation, that is, as having evolved by natural selection *in order to* perform a specific biological function in one or more environments. Speciationists called colleagues whom they thought were insouciant about meeting this burden of proof 'adaptationists'.[18] Self-declared adaptationists replied by saying that speciationists downplayed adaptive natural selection so much that they might not be Darwinians at all. By the 1980s this dispute had become so nasty and so public that it provoked doubts about the vaunted unity of the Synthesis, its continuity with Darwin's Darwinism, and its future viability as evolutionary biology's established theory.

Adaptationism has roots in suggestions by a group of religious thinkers at Oxford University in the early 20th century that adaptation by natural selection updates rather than contests Paley's argument from design. For Edward Poulton, Frederick Dixey and Aubrey Moore, natural selection is a law of nature that serves as vicar for God's creative act, a view still asserted by many religious evolutionists.[19] The Oxford school, shorn of natural theology, acquired authority in the middle decades of the century by using Fisher's approach to population genetics to confirm that in recent times dark-coloured peppered moths out-evolved lighter morphs for adaptive reasons, that Darwin's finches are adaptively differentiated by beaks sized and shaped by natural selection to efficiently utilise different food sources on different islands in the Galapagos chain, and similar cases.[20-22]

In the 1970s the adaptationist project acquired even greater caché when it used mathematical game theory, which it borrowed from economics and military strategy, to solve Darwinism's oldest and hardest problem: how a theory so committed to self-interest can explain the degree of cooperation and outright self-sacrificial behaviour that we see in nature, not least in our own species. Behavioural ecologists calculated that in the long run the optimal path of evolution for most species lies in a mixture of aggressive and accommodating behaviours: cooperative doves are essential if hawks are not to exterminate each other and hawks keep doves from failing to defend themselves and the population to which they belong. The Oxford behavioural ecologist Dawkins put this result on a more perspicuous footing when he argued that genes, now identified as stretches of DNA that remain intact through meiosis, can club together to evolve cooperative phenotypes in order to enhance their own reproductive success.[23] In this view, genes, not organisms, are the true beneficiaries of adaptive natural selection. By the 1980s widespread diffusion of behavioural evolutionary studies, aided by an effective publicity campaign applying the results to human evolution, conferred on genocentric versions of the adaptationist project a near monopoly on the term 'Darwinism' considered as a worldview. Speciationist theorems about reproductive barriers, species radiation, and the continuity of evolutionary mechanisms at all levels were accommodated by stressing their gradual adaptive causes. Since Darwinism so construed is resolutely naturalistic, even about ethics, the arc of successive Oxford schools shows ironic reversal from its theological beginnings.

To reinforce the point that adaptive natural selection works on a trait-by-trait basis, Dawkins conceived of organisms as assemblies of separately optimised parts, with no more integrity than 'clouds in the sky or dust storms in the desert'[23] (p. 34). In this atomistic ontology, organisms are sufficiently like well-designed machines, the parts of which are discrete, replaceable subunits, to cast natural selection as resulting in optimally good designs. I suspect that highly publicised efforts by advocates of this approach to beat back intelligent design spring ultimately from the uncomfortable fact that they share a conceptual framework in which organisms, whether designed by God or natural selection, are regarded as aggregates of independently adapted traits.[24] This point shows that by a conceptual frame I mean an ontology (and mereology) of a certain type. It also shows why I remarked that reductionist research strategies imply that organisms are not complex after all. In this framing they are no more than the sum of their parts.

The light of population thinking shone differently on the speciationist project. Following (or at least trusting in) the path-dependent mathematics of Sewall Wright, Dobzhansky and Mayr stressed the role of genetic drift in sparking genetic isolation, and hence speciation, in small, peripheral 'founder' populations.[25] Genetic drift is a phenomenon that cannot even be seen, let alone understood, without 'population thinking'. It means that in the short run, and in small populations, genotypes can spread for the same reason that a roulette player who observes the ball landing on red ten times in a row would be wrong to infer hastily either that the wheel is rigged or that black is bound to come up next time. After a large enough number of tries, red and black will come up in equal numbers, but a lot can happen by chance in the short run in small isolated populations.

Still, for Dobzhansky, and even for Wright, it was never the case that genetic drift leads to speciation without selection (*contra* Huxley).[26] 'Genetic drift never even in its meaning operates without selection', Dobzhansky told Mayr, 'so the notion of "fixation through genetic drift" is meaningless' (Dobzhansky to Mayr, October 18, 1954). Nor is it true that when Dobzhansky began stressing the role of adaptation in the speciation process in the late 1950s he was acceding to adaptationist arguments about Darwin's finches and dark-coloured peppered moths (*contra* Gould, p. 101).[27,28] Instead, he distinguished two kinds of adaptive natural selection. One adapts a species to a particular environment. The other evolves mechanisms that increase the likelihood of adaptability to changing environments, or what came to be called evolvability. The former can lead to evolutionary dead ends when environments change. Indeed, if adaptations for adapting had not evolved, such as diploid chromosomes that retain genetic variation in recessives and plasticity in the way the same genotypes respond to different environments ('norms of reaction'), evolutionary complexification would have ground to a halt long ago. Seen from this perspective natural selection probably retains, even maintains, in populations of peppered moths a balance between genotypes that respond differently to different circumstances (Dobzhansky[29] on balancing selection).

In the 1980s, speciationists responded to genic selectionism by arguing that variation and natural selection can simultaneously occur at various levels and at different rates in the hierarchical structuration of living systems.[27] Selection at the organismic level limits selection at the level of genes, which otherwise might be as non-adaptive as the runaway self-replication of cancerous cells. Buffering of this sort shows why organisms are hierarchically and modularly structured. By the same token, organism-level selection can be trumped by selection of and for traits in virtue of properties organisms have by being members of a reproductive group: group selection. The hierarchical model allowed David Sloan Wilson and Elliott Sober to demonstrate that applying game theory to evolutionary 'strategies' does not necessarily support the selfish-gene framework. Group-level selection (of a certain sort) allows cooperative behaviours to predominate in reproductive groups because, although destructive free riders certainly exist, they are often not powerful enough to overcome group solidarity — an idea Darwin himself intuited without having at his disposal any of the resources of mathematical game theory or the formal statistical reasoning that undergirds the Modern Synthesis.[30,31]

As adherents to the Synthesis, speciationists did not invoke development as an explanation of evolution, but neither did they entirely lose sight of the fact that natural selection has evolutionary import only when organisms are viewed developmentally. The dynamic responsiveness of genomes to environmental change over trans-generational time can be adaptive only if it is mediated by ontogeny in each generation. 'The organism is not specified by its genes', writes Lewontin, 'but is a unique outcome of an ontogenetic process that is contingent on the sequence of environments in which it occurs'[32] (p. 20). When genes are viewed as resources for development, the influence of one gene is always relative to the changing influence of others, so much so that Lewontin regarded the entire genome as the least unit of selection[33] (on genes as developmental resources[34,35]). In this spirit, following the lead of fellow Russian developmental geneticists Michael Lerner and Ivan Schmalhausen, Dobzhansky identified forms of natural selection that play key roles in the evolution of evolvability by representing aspects of development as trans-generational changes in gene frequencies, notably balancing, stabilising, normalising, and diversifying selection.[28,29]

In this approach, even so basic an idea as organism–environment interaction is meaningless outside of the developmental process. Environments are not external containers into which organisms are inserted or plugged in. Rather, in metabolically processing the resources from which they grow they exercise evolved abilities to construct their own environments and to evolve further by reacting to changes they themselves are continuously introducing.[32,36,37] They are agents in their own worlds, not passive products of environments that mould their traits. When people blithely partition the effects of genes and environments, especially in individuals, they are usually statically freezing the dynamic cycles of interaction in and through which development and evolution occur. In consequence they downplay the agency of organisms. These tendencies are encouraged by the artefact–organism analogy, but discouraged when genes and environments are seen in developmental interaction. To be sure, speciationists are adherents of the Modern Synthesis. They do not regard development as evolution in miniature. But that doesn't mean that they do not see organisms as ecologically embedded developmental systems.

The ontology of the speciationist project at its developmentalist best is most perspicuously framed in processual rather than substantial or thing-like terms. 'Our universal ancestor was not an entity but a process of genetic and energetic exchanges within protocellular communities'.[38–40] Organisms don't have life cycles. They *are* life cycles. 'The reality of metabolism forces us to recognize', write John Dupré and Daniel Nicholson, 'that despite their apparent fixity and solidity organisms are not material things but fluid processes'[41] (p. 13).[b] The perception of substantial solidity

---

[b]Dobzhansky retitled the last edition of his *Genetics and the Origin of Species* (1937, 1941, 1951) *Genetics of the Evolutionary Process*.[29] His denial that he was

that Aristotle was the first to stress and to contrast with non-substantial artefacts arises from misdescribed differences between the timescales of processes that unfold at different rates at different hierarchical levels of organic structuration. 'A cell persists far longer than any of its molecular constituents'[41] (p. 13). Similarly, an organism is longer lived than any of its cells, whose metabolic turnover is much more rapid. In turn, a species lasts longer than an organism, but not as long as the clade to which it belongs. Each level forms 'scaffolding' that allows evolution to take place at higher and lower levels[38,42] (p. 6).

It should be apparent that in recognising the role of conceptual frameworks I am not speaking ill of them. Rather than betraying empiricism they make it possible. Facts can't testify for or against hypotheses without a conceptual framework to identify the objects and processes they posit. Even to pose research questions fruitfully, conceptual frameworks are needed to organise and store the accumulated knowledge to which new discoveries might or might not be assimilated. It is also true, however, that those committed to a certain framework often believe that they are 'on the verge of making' new discoveries (about human psychology's relation to brain activity, for example) that in practice they assume have already been made. In doing so they risk flying off into a fact-free zone in which they appear not in the role of scientists but as devotees of a metaphysics, a theology, or an ideology to which they wish to convert the public and recruit their colleagues.[43]

In the 1950s and 1960s adaptationism was a progressive research program. Its practitioners made many discoveries about genes, traits, and environments. As long as it is treated defeasibly, the presumption that a trait is an adaptation is still useful in addressing a range of problems. Similarly, the speciationist program continues to facilitate new

---

a disciple of the process philosopher Alfred North Whitehead means that he regarded process as a conceptual framework for evolutionary biology, not an *a priori* metaphysics that imposes itself on and constrains empirical inquiry.

knowledge, even though its progress has depended on questionable assumptions: that evolution is change in gene frequencies, that adaptation and speciation are necessarily gradual and trans-generational, and that the evolution of higher taxa is smoothly continuous with known mechanisms of evolution at and below the species level.[44] Nonetheless, our growing knowledge of genetic and epigenetic regulation in the developmental process, to which I turn in the next section, seems to be undermining the picture of gene action that adaptationists and speciationists alike have assumed. At no time has the Synthesis been challenged more vigorously than in our own. It is even possible that its friends might soon have to choose between a metaphysically armoured Darwinian worldview and up-to-date biology. To opt for the first at the expense of the second is to opt for an *a priori* materialist metaphysics cut loose from scientific moorings.

The Future of Evolutionary Theory and the Revolution in Gene Regulation

Crick and Watson's discovery of the structure and function of DNA in 1953 was from the start expressed in a language of encoded, transmitted, received, and decoded information, a framing now so entrenched that we tend to forget that it is tropological.[45] Segments of DNA that complete this communication loop are said to contain all the information needed to make, even write a recipe for, an organism of a particular sort (or indeed a new, technologically begotten one). The trait-by-trait, gene-by-gene research program whose path I traced in the previous section relies heavily on the informational interpretation of DNA to recommend its key idea that genes code for traits that are assembled into organisms. In its 'selfish gene' formulation, traits are naturally selected 'vehicles' that interact with environments in order to maximise the more-making mechanism of DNA. In this framework engineering concepts of efficiency and optimality are essential in arguments applying this model to cases. Hence natural selection is conceived as design without a designer, a conceit that, as I have already remarked, is complicit enough with the intelligent design creationism it opposes to have provoked ardent protestations of atheism from defenders like Dawkins.[6]

It is clear from these considerations that genocentric versions of Darwinism draw heavily on the molecular revolution in genetics that got underway in the 1950s. But it is equally true that genocentrism gave molecular biology an evolutionary theory. It needed one. Molecular biology came out of biochemistry.[45–47] Its adepts had little awareness of the Modern Synthesis, the background of which is natural history, and to the extent that they knew about it they were typically indifferent or hostile to it. In the early, heady days of the molecular revolution, its practitioners sometimes called evolutionary naturalists 'stamp collectors', implying that they were clueless about the physical and chemical causes of evolution. Given this disciplinary divide, molecular biologists are often ill informed about the nuanced principles of the Modern Synthesis. They favour 'random muta-tion plus natural selection' as the canonical formula for Darwinian evolu-tion. Admittedly, this definition is widespread, in part because intelligent design theorists think it is an easy target to shoot at. But it lends itself to subtle misconceptions. It neglects gene flow and genetic drift, making adaptation a matter of definition rather than discovery and overlooking the interplay of mutation, selection, drift, and migration in speciation. The formula also implies that random mutation in DNA sequences is the only source of the variation on which natural selection feeds. At best, it is only the ultimate source. Moreover, in seeing variation as a precondition for natural selection rather than part of its definition, the formula can sug-gest that mutation is evolution's real driving force, sometimes in the form of fixation of variants by chance without any role for selection ('neutral mutation'). In consequence, it carries a suggestion that natural selection eliminates deviants from a norm rather than magnifying differences which, however defective they may be in one circumstance, can lead the way to adaptation and speciation in another.[48]

That is how natural selection looked *before* the Modern Synthesis. At that time the study of genes was dominated by eugenic thinking with racist overtones. It comes as no surprise, accordingly, that in offhand comments Watson has indulged in remarks that naively echo these themes. In 1994, for example, he told an interviewer that he was 'inherently gloomy about the prospect of Africa' because:

> all our social policies are based on the fact that their intelligence is the same as ours, whereas all the testing says not really ... I know that this 'hot potato' is going to be difficult to address. My hope is that everyone is equal, but people who have to deal with black employees find this not true .... There is no firm reason to anticipate that the intellectual capacities of peoples geographically separated in their evolution should prove to have evolved identically.[49]

Remarks like these are often taken as displays of Watson's bumptious character and ignorance of key facts, about human migration, for example.[c] He usually ended up apologising for them. Here I take them as betraying a conceptual deficit endemic in his disciplinary background. His successor as Director of the National Human Genome Research Institute, Francis Collins, shows no such character defects or knowledge gaps. In letting it be known that he is a believing and practising Christian, he has more or less immunised the NIH against the worries about hidden eugenic aims and racial prejudices that Watson's tenure encouraged. Still, in his public pronouncements on the ultimate aims of the Human Genome and other such projects, Collins construes natural selection as a designer commissioned by God somewhat in the spirit of the original Oxford School and intimates that, led by genetic biotechnicians, humans will increasingly be taking over as God's lieutenants in guiding the future of evolution on

---

[c]The science journalist Nicolas Wade has defended the claim Watson repeated about the adaptedness of races to different continents.[50] See Fuentes for a refutation of this assertion[51] and Jackson and Depew for a discussion of its historically and scientifically outdated context.[52]

Earth. Whether in its theistic or atheistic versions, I am far from sure that the design metaphor can bear this much weight. In consequence, I doubt whether either Watson or Collins can speak with confidence about the future of evolutionary science.

It was useful, indeed inevitable, that in the 1960s the effort to explicate the genetic code that ties the four bases of DNA to the twenty amino acids should have focused on structural gene products: proteins, which are composed of amino acids. By the end of that decade, however, it had become clear that point mutations in gene sequences for proteins, although they can certainly disrupt development, are less likely to be causes of evolution than changes in segments of DNA that, with help from enzymes, turn structural gene production on and off and speed it up or slow it down. First identified by the French geneticists André Lwoff, Francois Jacob, and Jacques Monod, regulatory genes have long been accommodated to the Modern Synthesis as genetically encoding what were already called 'norms of reaction'. They undergird, that is, a range of adaptive responses to a range of different environmental contingencies.

Increasingly, however, it appears that additional agents are at work in the interaction between development and environment. Genes make nothing, including themselves, without proteins. In a growing number of cases, short segments of RNA and epigenetic side chains of hydrogen and oxygen that attach themselves to DNA are also required if gene products are to be produced or blocked or their rate of protein synthesis is to be changed at a certain ontogenetic point.[53,54] These side chains are called *epigenetic* because they are *not* encoded in DNA or RNA.[d] In consequence, they are open to maternal influences. They are also heritable, if only for a few generations.[55] Should their significance continue to grow, development will no longer be viewed as exclusively a function of genotypes that have evolved a fixed, if wide, range of responses to various environmental

---

[d]Prions, which consist of (usually pathological) heritable variants, are also epigenetic by this standard.

conditions, as the Synthesis would have it, but instead as a process in which developmental resources of many kinds, some under environmental influence, are continuously interacting.

If that happens, we will be witnessing a phenomenon first described by Kuhn in *The Structure of Scientific Revolutions*. Anomalies that have long been swept under the rug by the ascendancy and disciplinary imperatives of the Modern Synthesis will suddenly become salient and explicable. Among these are what the pioneering developmental geneticist C. H. Waddington (inaccurately) called 'genetic assimilation', in which environmental effects become genetic, and cases of adaptation and speciation that occur too rapidly to fit comfortably within the conceptual boundaries of the Synthesis.[52] It is not even true that genes always code for the same proteins. Alternative splicing of DNA occurs all the time. Could this process, too, be responsive to environmental factors?[e] Some who have chafed under the domination of the ruling paradigm are already predicting a renaissance of Lamarckism, which before and certainly after the molecular revolution was anathema to the makers of the Modern Synthesis.[55] This prospect has already been stimulating worries that a new version of the heritability of acquired characteristics will revive social prejudices that genetic Darwinism, backed by the molecular biology, succeeded in discrediting.[f]

At this point we can appreciate yet another irony in the history of evolutionary biology. Efforts to map and sequence whole genomes, especially the Human Genome Project, were at first predicated on the assumption

---

[e]The fact that DNA is open to insertion of other bits of DNA does not contradict the letter of Crick and Watson's Central Dogma of Molecular Biology, according to which information flows from DNA to RNA and proteins but not the other way around, but it does conflict with its spirit. These insertions are very common in nature. They are the foundation of modern genetic biotechnology, most recently the CRISPR-Cas 9 process of incorporating DNA sequences into virtually any part of any genome.

[f]'Epigenetic explanations can tap into longstanding debates on the reproduction of poverty and may invigorate them by offering a visible mechanism for the biological perpetuation of race and class pathologies'[56] (p. 222).

that genes encode the destiny of each individual. All of our traits, we were told, are specified in our unique genomes. Once our genetic risks are established the search for treatments could systematically get under way.[57] It came as something of a surprise, accordingly, that the research in question showed that the number of human genes falls far below estimates and that most of our genes are shared with taxa stretching back as far as the first microbes. At first, this realisation did not call into question the assumptions it confuted. Instead, money to find the missing information began to flow into 'proteinomics' and 'epigenomics'. It was in this context that epigenetics came to the fore, and with it challenges to the utopian, but simplistic hopes that gave rise to the Human Genome Project in the first place, and to genetic determinist versions of the Modern Synthesis that I summarised in the previous paragraphs.

One response to this situation has been the project called 'systems biology'. The idea is to collect 'big data' about every level of biological activity and to use advanced computation to find predictive correlations between them. Among these levels will be genomic, cellular, organismic, and ecological factors, the last of which, far from being a dependent variable of lower level processes, has a dynamics of its own.[58] It would be unfortunate if systems biologists took themselves to be mastering complexity merely by finding such correlations. Correlating with a view to prediction and biotechnological manipulation may pay practical dividends, but from the perspective of advancing evolutionary theory it strikes me as missing an opportunity to use new mathematical means to recognise the multi-level complexity of organisms, their coevolutionary relationship with environments, and their nature as developmental processes. As I remarked earlier, it is one thing to appreciate the complexity of living things and quite another to reduce it to simplicity. A better approach is to look for mathematical means of capturing the dynamics of irreducibly complex systems. As it happens, these tools have been coming into use for some time, especially in the form of network models that identify dynamic interactions between the various nodes of a system, weight their interactions, and track their changes over time.[38]

Conclusion: Innovation and Tradition in Evolutionary Biology

In Kuhn's seminal model of scientific change and its more refined successors, a new paradigm or research tradition will displace an old one only if it preserves most of the solid results of its predecessors by recasting them in its terms and rederiving them from its principles.[59–61] What fails to survive this sifting will no longer be recognised as scientific knowledge. In just this way, astrology was subtracted from astronomy, alchemy from chemistry, and, after Darwin, creationism from the study of adaptation. A similar sorting is now underway that pits evolutionary developmentalism ('evo-devo') against the 'population thinking' of the Modern Evolutionary Synthesis. Where this struggle will end is difficult to predict.

Developmentalist challenges to the Modern Synthesis are especially intense when both its apologists and opponents construe Darwinism as a materialist worldview based on trait-by-trait adaptationism. When this becomes the meaning of 'Darwinism' its enemies are quick to claim that the evolutionary theory that arose in Darwin's wake misses the proper explanation of evolutionary phenomena because more fundamentally it misses what an evolutionary theory must explain: the architecture of the forms identified by comparative anatomy. It is an old accusation. Richard Owen raised it against Darwin himself. Owen's contemporary avatars believe that variations on the fundamental body plans on which systematic classification is based are now explainable by sudden switching in highly conserved regulatory genes.

If a dichotomy this severe were to gain ascendancy, the assumption of the makers of the Modern Synthesis that macroevolution is a long-term consequence of microevolution would be abandoned. In consequence, the vast knowledge of microevolution — evolution at and below the species level — that has been accumulating for a century would be devalued and hard-fought efforts to make the study of evolution a science would be threatened. Creationists would have something to crow about. In part because of these unpalatable, and in fact unlikely, implications I have taken pains to distinguish rather sharply between adaptationist and speciationist research programs in the Modern Synthesis and to argue that there are developmentalist themes in the speciationist tradition that can be recast in a new, emerging paradigm that expands and may eventually replace the Modern Synthesis.[52] Most proposals to expand the Synthesis reconceive the explanatory power and unity of evolutionary biology by casting developmental changes as generating the proximate variation on which natural selection works to build changed architectures, sometimes in only a few generations, and accord greater agency to organisms.[62,63]

The proposed shift integrates evolutionary developmentalism with what the Modern Synthesis has already learned by rejecting the conceptual assumptions that have long bound its speciationist and adaptationist wings. Both orientations have taken it as axiomatic that adaptation is necessarily trans-generational because genomes adjust to environmental changes only gradually. This assumption has become a fetter on evolutionary science.[52] Reformulating the study of evolution in terms of process ontology will, I believe, liberate the submerged developmentalism of speciationists. It will also lead to better interpretations of what we know about adaptation by freeing it from design thinking, which in seeing organisms as passive things rather than active processes misses development as the locus of evolution.[9] It would be helpful if science lecturers, writers, and teachers would suggest that contemporary evolutionary thought is exciting because it is giving the living world back to us in understanding, not stripping it of its enduring capacity to inspire awe.

---

[9] Hegel recognised 'thing' talk as a degenerate post-medieval version of substance ontology in which, in contrast to Aristotle's view, artefacts and organisms have the same ontological status. For Descartes and Kant, even God is a thing, an *ens realissimum*. Whether process ontology, in addition to serving as a conceptual framework for the science of evolving organisms, is also a general science of being *qua* being, a metaphysics, is a question that depends in part on whether it can do justice to our intuitions about persons. In spite of, or perhaps because of, rapid increase in empirical knowledge about the brain, these intuitions must still be protected by philosophical resistance to reductionist and eliminationist conceptual framings. This situation of inquiry is likely to last for a long time.

References

1. Pauling L. (1939) *The Nature of the Chemical Bond and the Structure of Molecules and Crystals.* Cornell University Press, Ithaca, New York.
2. Watson JD. (1968) *The Double Helix.* Atheneum, New York.
3. Watson JD and Crick FHC. (1953) Molecular structure of nucleic acids: A structure for deoxyribose nucleic acid. *Nature* **171**: 737–738.
4. Wilson EO. (1998) *Consilience: The Unity of Knowledge.* Knopf, New York.
5. Kuhn TS. (1962) *The Structure of Scientific Revolution.* University of Chicago Press, Chicago.
6. Dawkins R. (2006) *The God Delusion.* Houghton Mifflin, Boston.
7. Gould SJ. (1977) *Ontogeny and Phylogeny.* Harvard University Press, Cambridge, MA.
8. Fisher RA. (1930) *The Genetic Theory of Natural Selection.* Oxford University Press, Oxford.
9. Provine W. (1971) *The Origins of Theoretical Population Biology.* University of Chicago Press, Chicago.
10. Weldon WFR. (1893) On certain correlated variations in *Carcinus moenas. Proc R Soc* London **54**: 318–329.
11. Dobzhansky T. (1937) *Genetics and the Origin of Species.* Columbia University Press, New York.
12. Gilbert S. (1994) Dobzhansky, Waddington, and Schmalhausen: Embryology and the Modern Synthesis. In: Adams M (ed.) *The Evolution of Theodosius Dobzhansky,* pp. 143–154. Princeton University Press, Princeton.
13. Gilbert S and Epel D. (2008) *Ecological Developmental Biology: Integrating Epigenetics, Medicine, and Evolution.* Sinauer Associates, Sunderland.
14. Mayr E and Provine W. (1980) *The Evolutionary Synthesis.* Harvard University Press, Cambridge, MA.
15. Smokovitis VB. (1996) *Unifying Biology: The Evolutionary Synthesis and Evolutionary Biology.* Princeton University Press, Princeton.
16. Jackson J and Depew D. (2017) *Darwinism, Democracy and Race.* Routledge, London.
17. Lewontin RC, Paul D, Beatty J and Krimbas C. (2001) Interview of R. C. Lewontin. In: Singh R, Krimbas C, Paul D and Beatty J (eds.) *Thinking About Evolution: Historical, Philosophical and Political Perspectives: A Festschrift for Richard C. Lewontin,* pp. 22–61. Cambridge University Press, Cambridge.
18. Gould SJ and Lewontin RC. (1979) The spandrels of San Marco and the Panglossian paradigm: A critique of the adaptationist program. *Proc R Soc B-Biol Sci* **205**(1161): 581–598.

19. England R. (2001) Natural selection, teleology, and the *logos*: From Darwin to the Oxford neo-Darwinians. *Osiris* **16**: 270–287.

20. Lack D. (1947) *Darwin's Finches*. Cambridge University Press, Cambridge.

21. Kettlewell HBD. (1955) Selection experiments on industrial melanism in the *Lepidoptera*. *Heredity* **9**(3): 323–342.

22. Cain AJ, King JMB and Sheppard PM. (1960) New data on the genetics of polymorphism in the snail *Cepaea nemoralis*. *Genetics* **45**: 393–411.

23. Dawkins R. (1976) *The Selfish Gene*. Oxford University Press.

24. Behe M. (1996) *Darwin's Black Box*. Free Press, New York.

25. Wright S. (1932) The roles of mutation, inbreeding, crossbreeding and selection in evolution. *Proc 6th Int Cong Genet* **1**: 356–366.

26. Huxley J. (1942) *Evolution: The Modern Synthesis*. Allen and Unwin, London.

27. Gould SJ. (1983) The hardening of the Modern Synthesis. In: Grene M (ed.) *Dimensions of Darwinism*, pp. 71–93. Cambridge University Press, Cambridge.

28. Depew D. (2008) Consequence etiology and biological teleology in Aristotle and Darwin. *Stud Hist Philos Biol Biomed Sci* **38**(4): 379–390.

29. Dobzhansky T. (1970) *Genetics of the Evolutionary Process*. Columbia University Press, New York.

30. Sober E and Wilson DS. (1998) *Unto Others: The Evolution of Altruism*. Harvard University Press, Cambridge, MA.

31. Darwin C. (1871) *The Descent of Man*. Chapman, London.

32. Lewontin RC. (2000) *The Triple Helix: Gene, Organism, and Environment*. Harvard University Press, Cambridge, MA.

33. Lewontin RC. (1974) *The Genetics of Evolutionary Change*. Columbia University Press, New York.

34. Griffiths P and Gray R. (1995) Developmental systems and evolutionary explanation. *J Philos* **91**: 277–304.

35. Moss L. (2003) *What Genes Can't Do*. MIT Press, Cambridge, MA.

36. Lewontin RC. (1982) Organism and environment. In: Plotkin H (ed.) *Learning, Development and Culture: Essays in Evolutionary Epistemology*, pp. 151–170. John Wiley and Sons, New York.

37. Odling-Smee F, Laland K and Feldman M. (2003) *Niche Construction: The Neglected Process in Evolution*. Princeton University Press, Princeton.

38. Bapteste E and Huneman P. (2018) Toward a dynamic interaction network of life to unify and expand evolutionary theory. *BMC Biol* **16**: 56 http://doi.org/10.1186/s12915-018-0531-6.

39. Woese C. (1998) The universal ancestor. *Proc Natl Acad Sci USA* **95**: 6854–6859.

40. Weber BH. (2013) Complex systems dynamics in evolution and emergent processes. In: Henning B and Scarfe A (eds.), *Beyond Mechanism: Putting Life Back into Biology*, pp. 67–74. Lexington, Lanham, MD.

41. Dupré J and Nicholson DJ. (2018) Toward a processual philosophy of biology. In: Nicholson DJ and Dupré J (eds.). *Everything Flows*. Oxford University Press, Oxford.

42. Caporael L, Griesmer J and Wimsatt W. (2013) *Scaffolding in Evolution, Culture, and Cognition*. MIT Press, Cambridge, MA.

43. Depew D. (2013) The rhetoric of evolutionary theory. *Biol Theory* **7**(4): 380–389.

44. Gould SJ. (1980) Is a new and general theory of evolution emerging? *Paleobiology* **6**: 119–130.

45. Kay L. (2000) *Who Wrote the Book of Life? A History of the Genetic Code*. Stanford University Press, Stanford, CA.

46. Dietrich M. (1994) The origins of the neutral theory of molecular revolution. *J Hist Biol* **27**: 21–50.

47. Judson HF. (1996) *The Eighth Day of Creation*. Commemorative ed. Cold Springs Harbor Laboratory Press, New York.

48. Depew D. (2015) Myth 20: That neo-Darwinism is random variation plus natural selection. In: Numbers RL and Kampourakis K (eds.) *Newton's Apple and Other Myths about Science*, pp. 164–170. Harvard University Press, Cambridge, MA.

49. Hunt-Grubbe C. (2007) The elementary DNA of Dr. Watson. *The Times of London,* October 14, 2007.

50. Wade N. (2015) *A Troublesome Inheritance*. Penguin Books, New York.

51. Fuentes A. (2016) *Race, Monogamy, and Other Lies They Told You: Busting Myths About Human Nature*. Berkeley and Los Angeles: University of California Press.

52. Depew D and Weber BH. (2017) Developmental biology, natural selection, and the conceptual boundaries of the Modern Evolutionary Synthesis. *Zygon* **52**: 468–449.

53. Wan G, Fields B, Spracklin G, *et al.* (2018) Spatiotemporal regulation of liquid-like condensates in epigenetic inheritance. *Nature* **557**: 679–683.

54. Gadjev I. *Homo faber*, Will, Determinism, and Heredity: From Genetics to Epigenetics. *(This volume.)*

55. Jablonka E and Lamb ML. (2005) *Evolution in Four Dimensions*. MIT Press, Cambridge, MA.

56. Meloni M. (2015) *Political Biology: Science and Social Values in Human Heredity from Eugenics to Epigenetics*. Palgrave Macmillan, Hampshire.

57. Kevles D and Hood L (eds.). (1994) *The Code of Codes: Scientific and Social Issues in the Human Genome Project*. Harvard University Press, Cambridge MA.

58. Gunton R and Gilbert F. The Reintegration of Biology, or 'Nothing in Evolution Makes Sense Except in the Light of Ecology'. *(This volume.)*

59. Kuhn TS. (1970) *The Structure of Scientific Revolutions*, 2nd edn,. University of Chicago Press, Chicago.

60. Lakatos I. (1968) Criticism and the methodology of scientific research programmes. *Proc Aristotelian Soc* **69**(1): 149–186.

61. Laudan L. (1977) *Progress and Its Problems*. University of California Press, Berkeley, CA.

62. Pigliucci M and Muller GB (eds.). (2010) *Evolution: The Extended Synthesis*. MIT Press, Cambridge, MA.

63. Huneman P and Walsh DM (eds.). (2017) *Challenging the Modern Synthesis — Adaptation, Development, and Inheritance*. Oxford University Press, Oxford.

Human Evolution: From Fossils to Molecules, Reductionism to Holism

Elizabeth Jones* and Michael Ruse[†,‡]

'Ein Volk, Ein Reich, Ein Führer' Motto of Nazi Germany

Our Questions

'It is a truth universally acknowledged, that a single man in possession of a good fortune, must be in want of a wife'. It is also a truth universally acknowledged that anyone who presumes to write on human nature is entering a minefield. It is, to put it mildly, a much-contested topic, especially if one presumes to write on biological aspects of human nature. We do so presume, however, and we are going to ask a question that many would think needs little thought to answer. Dividing scientific approaches somewhat crudely into those that take a 'reductionistic' approach — always seeking explanations in terms of yet-smaller entities — and those that take a 'holistic' approach — allowing, insisting on, explanations that stress the unity of the

*History of Science Research Lab, North Carolina Museum of Natural Sciences, Raleigh, NC 27601, USA.
†Department of Philosophy, Florida State University, Tallahassee, FL 32306, USA.
‡Professor Emeritus, University of Guelph, Canada.

system or process — where do we find ourselves with respect to humankind? Perhaps, where *should* we find ourselves with respect to humankind? The obvious answer — the morally obvious answer — is that we must and should stay away from holistic understandings. That way lies National Socialism.[a]

Prima facie, although understandable, this reaction is somewhat odd. Usually 'holism' or 'holistic' is taken to be a word of praise. Holistic therapy, holistic agriculture, holistic living. Even holistic dog food! 'The WHOLE difference. Compare: 100% WHOLE meat, poultry, or fish. Never ANY rendered "Meat", like "chicken meal" or "fish meal"'. Comforting, cuddly, Californian. And yet, here we are being told to veer right away, and to have nothing to do with such holism. Something interesting is going on here, and as human fools rushing in where angel purists fear to tread, we are going to ask about human nature and the holism/reductionism question. More specifically, we are going to ask about our understanding of human evolution and the holism/reductionism question.

Although this is our central question, it is yet another universally acknowledged truth that there is no such thing as a private science of human nature, of human evolution in particular. This started as soon as the Origin was published. The humorous magazine *Punch*, for instance, carried non-stop cartoons linking Darwin with the apes. Often, turning him into a kind of primate centaur. The same is true today. Before anything interesting has made its appearance in the top journals like *Science* and *Nature*, you can be sure that it has been expounded and discussed in the *New York Times* and the *Guardian*. It has even been known to make its way onto Fox News, although both authors rush to say that they know this only by hearsay. Be this as it may, one secondary question that is at the back of our

---

[a] The holistic philosophy of the Nazis — one people, one state, one leader — was nothing new in German political thinking. It goes back at least to the thinking of the early nineteenth century philosopher Georg Friedrich Wilhelm Hegel, whose philosophy of Absolute Spirit lent itself readily to nationalistic aspirations. After the unification of Germany in 1871, such political holism was pushed strongly, especially by the Prussian military who were hugely influential. This persisted after the First World War and meshed happily with the National Socialist philosophy. Note what a significant role the military played in the Third Reich.

minds is whether the public nature of the debate feeds into the science and has any effect on the holism/reductionism question. Does the very fact of being public shape the scientific discussion, perhaps making people less willing to make certain assertions than otherwise? If you feared for instance that a thousand people would be all over you on Facebook, would you make those allegedly certain assertions quite so confidently and publicly?

Human Evolution and the Fossil Record

In the eighteenth century, as more and more of the globe was visited and explored, increasingly there came back to Europe reports of strange but vaguely human-like beasts, Pongos and Jockos and more. Two species did claim serious attention, the chimpanzees and the orangutans, so much so that when the Swedish systematist Linnaeus came to classify our own species, *Homo sapiens*, he included these two other species in the same genus, *Homo*, all part of the primates. Linnaeus was no evolutionist, but it was (one might say) the thin end of a very large wedge. It was not long before people were starting to speculate about origins. What we now call 'evolution' was the obsession of many, including the grandfather of Charles Darwin, the British physician Erasmus Darwin. Often expressing his ideas in poetry, he shows nicely that the real interest in evolution was in humankind. No one cared much about warthogs. Did we humans evolve and if so when and where and from what?

> Imperious man, who rules the bestial crowd,
> Of language, reason, and reflection proud,
> With brow erect who scorns this earthy sod,
> And styles himself the image of his God;
> Arose from rudiments of form and sense,
> An embryon point, or microscopic ens![1] (p. 15)

It was Charles Darwin who made all of this into a science, showing that all life including humans is the end product of a slow process of change and development, from just few primitive forms, by a mechanism that he labelled 'natural selection'. More organisms are born than can survive and reproduce. Those who are successful are on average different from those that do not. Thanks to an ongoing introduction of new variation into populations, there will be ongoing winnowing and change. Moreover, this change is in the direction of being design-like, helping their possessors. Adaptations like the eye and the hand. 'Can we wonder, then, that nature's productions should be far "truer" in character than man's productions; that they should be infinitely better adapted to the most complex conditions of life, and should plainly bear the stamp of far higher workmanship?'[2] (pp. 83–84)

As it happens, in his *Origin of Species*, published in 1859, where he expounded his theory, Darwin said virtually nothing about our own species. In the most understated comment of the nineteenth century he opined: 'Light will be thrown on the origin of man and his history'[2] (p. 488). He said this lest he be accused of cowardice. He knew that immediately people would be asking about human beings, and he wanted first to get his theory on the table as it were. Darwin's forebodings were right.[b] Darwinism became the sensation of the age and *Homo sapiens* and its place in the monkey or gorilla theory was the question of the hour. The Bishop of Oxford, Samuel Wilberforce, asked the Professor at the School of Mines, Thomas Henry Huxley, if he was descended from monkeys on his father's side or his mother's side. To which the professor responded better a monkey than a bishop of the Church of England. Prime Minister Benjamin Disraeli joked that he had rather be an angel than an ape, not the first time that a conservative premier has made the wrong decision.

From the first, Darwin was stone-cold certain that we are fancy apes — what Huxley was to call modified monkeys rather than modified dirt. In

---

[b] Darwin's book and theory of evolution by natural selection initiated comments and concerns from some his closest collaborators and critics, particularly as the theory related to religion — Christianity specifically. While the theory presented some tough questions that many found difficult to fully reconcile, at least at first, Darwin's close circle of friends — from Alfred Russel Wallace, Joseph Dalton Hooker, and Asa Gray to name a few — did not necessarily find that evolutionary theory and faith in a higher deity were incompatible. See Dobson[3] for more information.

his private notebooks, in the first record we have of Darwin using natural selection, he applies it not just to humans but to our intelligence:

> An habitual action must some way affect the brain in a manner which can be transmitted. — this is analogous to a blacksmith having children with strong arms. — The other principle of those children, which chance? Produced with strong arms, outliving the weaker ones, may be applicable to the formation of instincts, independently of habits.[4] (N 41, Nov 27th 1838)

After the *Origin*, articles and books on our species began to pour forth. Most notable was *Man's Place in Nature* by Thomas Henry Huxley. Although always doubtful of the full power of natural selection, Huxley made it very clear that we are part of the natural order and that we are indeed descended from apes — not necessarily apes living today, but apes nevertheless. Already in the late 1820s Neanderthal fossils were being unearthed (they are named after a valley in Germany) and Huxley naturally speculated on their relationship to us. He decided that they are a subspecies of *Homo sapiens*, unlike others who made Neanderthals entirely separate. Thus began the long search for an answer to the question of whether they ever interbred with us modern humans. The older of the two authors remembers Sir Karl Popper solemnly assuring him that this was quite impossible. Not all were of this opinion, and this applies not only to those overexposed to Neanderthal-like teenagers. There were those who inclined to think that thriving, interbreeding Neanderthals could be found on the west coast of Ireland.

Darwin never really enjoyed controversy and probably he would have stayed out of the searchlight. He had done enough for one scientific lifetime! This was not to be. In the 1860s, the co-discoverer of natural selection, the naturalist Alfred Russel Wallace, took up spiritualism, and argued that human evolution demanded spirit forces to achieve its ends. Appalled, Darwin countered with the *Descent of Man* published in 1871, where he argued the case for our natural origins. He did give Wallace one major point. The latter had argued that certain human features — hairlessness, big brains and intelligence for instance — could not have been produced by natural selection. Agreeing, Darwin made much of his secondary mechanism of sexual selection, arguing that the features highlighted by Wallace could have been produced naturally in the struggle for mates. No need for spiritualism! (Darwin was never an atheist, but he did want his science to be science, meaning a science like the physical sciences.)

What about the all-important question of the 'missing link'? Although he started his scientific life as a geologist, Darwin was never really a palaeontologist. He was more a secondary figure here, starting with the debate about where to look. Darwin favoured Africa for the birthplace of humans, but this was not a popular option. No self-respecting European liked the idea of coming from Africans. Asia was the Number One choice, and this seemed vindicated in the early 1890s when the Dutch physician Eugene Dubois won the race. He found, 'Java Man', whom he put in a new genus *Pithecanthropus erectus*. More recently, it has been reassigned to our genus *Homo erectus*. It was a genuine missing link. We have brains of around 1200 cc, chimpanzees are down around 400cc, and Java Man came in at 900cc.

Study of human evolution — 'palaeoanthropology' — was thrown off course in the second decade of the last century by the greatest fraud in the history of science — Piltdown Man. Discovered, or more accurately 'discovered', in the South of England, supposedly we had a being with a human-like cranium and an orang-like jaw. It was in the 1950s that it was realised that the reason why the parts are human-like and orang-like is because they are in fact human and orang. By then the damage had been done, misleading research, as much because people were so happy to find links in Britain (as opposed to the German Neanderthals) as that the evidence was truly convincing. One subsequent find had a bone carved in the shape of a cricket bat!

Science is like the Hegelian world force. You can disrupt it for a while, but history has its aims. Slowly but surely Africa started to take its place as our birthplace. A crucial discovery was 'Taung Baby', discovered in South Africa in 1924 and championed by (yet another medical man) Raymond

Dart. *Australopithecus africanus* really is in a new genus, having a brain size of about that of a chimpanzee, 400cc (or a bit more), although we stress that the internal indentations of the cranium show that it was not a chimp brain but one on the way to us humans. The relevance of the find was strongly disputed, especially by the British establishment, who were firmly wedded to Piltdown and the pre-eminence of anything British.[c] However, more African fossils started to appear and, in the 1950s, the indefatigable Leakey family — Louis and Mary and children including Richard — started work in East Africa. The Leakeys soon started to reveal a treasure house — another species of australopithecine (Zinj, now put in *Australopithecus boisei*) as well as more specimens of *Homo*. Perhaps the most famous of all these discoveries is associated with the American palaeoanthropologist Donald Johanson. Lucy, now classified as *Australopithecus afarensis*, is a wonderful specimen of a hominin (ancestor of humans), with an ape-sized brain but clearly bipedal. The missing link indeed, and feted in newspapers, magazines, and popular books, as well as in the professional publications. Although unbelievably self-promoting — actually, not so unbelievable once you grasp the kinds of personalities attracted to this kind of work — the book (*Lucy*) that Johanson co-authored with a professional writer is a terrific read.

In major respects, the human fossil lineage is very well represented, and continues to grow and to throw up major surprises. Most recently, about fifteen years ago was the finding of a strange little creature, about three and a half feet tall — *Homo floresiensis* — at once, perhaps inevitably labelled the 'hobbit' (and, perhaps as inevitably, a use challenged legally by the Tolkien estate). Likewise controversial — persistent critics claimed it was nothing but disabled human — the hobbit hailed from one of the

---

[c] It is important to note that it has been argued that one of the major reasons why Dart's fossil find was so contested was because he refused to ship the specimen to London or any other city centre for comparison with other specimens by other researchers in his field. In refusing to send the specimen for comparison, Dart defied the scientific expectations of research at the time and his career and his ideas suffered from his unconventional attitude towards conventional scientific practice as a result.

islands of Indonesia, Flores. But, although controversial, no one denied that the existence of a being like this was fully consistent with Darwin's theory of evolution through natural selection.

Thus, the story of the fossils. In a way, all rather old-fashioned and Victorian, starting with the fact that, from the first, every thought and every act was done in the public footlight. Nothing that would have phased those old rivals, Richard Owen and Thomas Henry Huxley — for both of them, there was no such thing as too much publicity. There was new material of course, but there were the same old techniques, even if refined. Nothing very much to do with the holism/reductionism debate. Anthropologists looking at today's humans might have their issues, but not the palaeoan-thropologists.

A bombshell was thrown smashing this happy picture — perhaps, more accurately, this happy complacency. In 1953, James Watson and Francis Crick (not to mention the at-the-time-unsung Rosalind Franklin) discovered the double helix, the nature of the carrier of the information for life, the deoxyribonucleic acid, DNA. Initial reaction by conventional evolutionists was expectable. They denied the relevance of the molecular approach. Ernst Mayr, well-known systematist and ornithologist, went so far as to say that such an approach is grotesquely reductionistic and can tell us nothing of importance about human evolution! How very wrong he was. The science of human origins was about to become the most reductionistic science the world had ever known and even more exciting than it had ever been.

Ancient DNA and the Molecular Fossil Record

This past spring, David Reich — one of the foremost geneticists behind the search for DNA from human fossils — published the first comprehensive work on the study of the origins, evolution, and migration of the human species as told with new evidence from the new science of ancient DNA research. Unlike Darwin's understatement of the century at the end of his big book on The Origin of Species — 'Light will be thrown on the origin of man and his history' — Reich's book tackles the topic head first. Since the 1800s, the scientific search into our human origins had been limited to the study of fossil human remains. In the 1950s, however, the discovery of the structure of DNA changed the game, and much more recently, the discovery of ancient genetic and genomic information from some of the world's oldest fossils has changed our view of the past as well as our place within it. In the introduction, Reich claimed that 'Ancient DNA and the genome revolution can now answer a previously unresolved question about the deep past: the question of *what happened* — how ancient peoples related to each other and how migrations contributed to the changes evident in the archaeological record'[5] (p. xxiv). Reich warmly welcomed this new line of evidence, remarking that archaeologists should be equally excited by the prospect of ancient genetic and genomic data: 'Ancient DNA should be liberating to archaeologists because with answers to these questions in reach, archaeologists can get on with investigating what they have always been interested in, which is *why* these changes occurred'[5] (p. xxiv). According to Reich, ancient DNA research does more than inform our understanding of human history: he says it has transformed, and will continue to revolutionise, our understanding of who we are, how we got here, and how we relate to one another today.

'Ancient DNA Research' — the practice of extracting, sequencing, and analysing degraded or damaged DNA from dead organisms that are hundreds to thousands of years old — is a practice that first emerged in the 1980s from the interplay between palaeontology, archaeology, and molecular biology.[d] Over the decades, DNA has been recovered from

---

[d]The search for DNA from ancient and extinct organisms is an outgrowth of years of scientific and technological changes and an even longer history of conceptual and cultural evolvements. In the latter half of the twentieth century, new areas of scientific study were born out of important innovations from electron microscopy to protein sequencing and recombinant DNA research. The birth of ancient DNA research is yet another one of these new areas of inquiry whose growth is part of a contingent and contentious process.[6–19] For a specific history of ancient DNA research.[20–24] For a technical discussion of ancient DNA research.[25]

ancient and extinct organisms like plants, animals, humans, and even bacteria.[e] It can be preserved in skins, tissues, and even bone if the bone is not a fully mineralised fossil.[f] Today, many scientists are mainly engaged in adapting state-of-the-art molecular biological techniques and high-throughput sequencing technologies to enhance the recovery of DNA from ancient material.[g] Broadly, their goal is to use this molecular information in order to test hypotheses about evolutionary processes, including testing for mechanisms that drive and develop patterns of genetic variation, discovering regions of the genome under selection, and understanding migrations of past populations.

The science of ancient DNA research, much like the science of human evolution, has always been a public-facing practice. From the beginning, the search for DNA from fossils emerged and evolved in the public spotlight, especially as its first years conveniently corresponded with the release of Michael Crichton's book (1990) and movie (1993) *Jurassic Park*.[21,22] In the 1990s, the hunt for DNA from fossils was transformed into a high-profile

---

[e]Note that the term 'ancient' does not necessarily relate to the age of the DNA but to the characteristic damage patterns that occur as DNA breaks down after an organism has died. Given the degraded nature of this genetic material, research into DNA from ancient and extinct species requires specialist skills and technologies.

[f]Specifically, a fully mineralised fossil is unlikely to preserve DNA, whereas a sub-fossil, a partially mineralised part of an organism, may retain remains of its cellular or molecular components. Specifically, an organism's status as a fossil or subfossil, or whether it exists as a piece of skin or tissue, matters when considering whether cellular or molecular components may be preserved.

[g]Next generation sequencing (NGS) is the general term used to describe a variety of technologies that use parallelised platforms to sequence more than one million short reads of DNA (50–400 base pairs) in a single run. There are a number of NGS platforms varying in their chemistry and specific sequence read technologies. Two instruments that were widely used in ancient DNA research in the late 2000s were Roche (454) GS FLX, a technology based on parallel pyrosequencing, and Illumina (Solexa) Genome Analyzer, a method based on reversible termina-tors. The 454 technology generates longer reads of DNA (over 400 base pairs) but is somewhat error-prone in homopolymeric regions, while Illumina generates shorter reads of DNA (100–150 base pairs), but in greater numbers.[26,27]

science and technology as a series of studies, published in high-impact journals such as *Nature* and *Science,* reported the recovery of multi-million year old DNA from fossils such as amber insects and even dinosaur bone.[28-32,h] Riding on the heels and hype of *Jurassic Park,* researchers raced to extract and sequence DNA from the days of the dinosaurs. In the process, the press created opportunities for publicity, but scientists also fashioned their own opportunities for attention. During this decade, scientists were savvy in capitalising on the celebrity of their fast-growing field in order to secure their success on both an individual and group level. Here, the interplay between science and the media, specifically around the idea of discovering multi-million year old DNA, influenced research agendas, publication timing, grant funding, and public perceptions of the search for DNA from fossils. Ancient DNA was far from a private affair. Indeed, it was, and still is, a public one. Sure enough, the field continues to capture professional and popular attention as researchers have recovered genetic material from extinct mammoths to early humans and Neanderthals in an attempt to refine or even rewrite our understanding of evolutionary history.

In 1997, two teams of scientists — one in Germany and one in the United States — reported the recovery of Neanderthal DNA, mitochondrial DNA (mtDNA) specifically.[33] The Neanderthal mtDNA when compared to mtDNA of primates and modern humans from Africa, Europe, Asia, and across the world demonstrated that their differences (based on a single sequence) were distinct. Consequently, the researchers interpreted this as evidence that Neanderthals had lived, then died, without contributing any of their DNA to modern humans. These results also suggested that modern humans had their origin in Africa not Europe. But researchers reasoned that these research results did not entirely eliminate the possibility of a genetic contribution from extinct Neanderthals to extant humans. Nonetheless, this study and its attempt to use molecular data to inform a history that had traditionally relied exclusively on morphology added heat to an already

---

[h]As it happened, these studies were later shown to be the result of contamination.

heated debate in evolutionary anthropology about our own origins and evolution over time.[i]

In 2006, nearly ten years after these first findings, Svante Pääbo — one of the founders of ancient DNA research — announced at a press conference and with a press release that his lab at the Max Planck Institute for Evolutionary Anthropology and 454 Life Sciences Corporation would be the first to sequence the entire Neanderthal genome — and they announced they would do it in just two years' time.[35,36,j] The Neanderthal Genome Project, although an ambitious and one of a kind project on its own, was not necessarily an isolated idea, particularly as it rode a wave of research, like the Human Genome Project, interested in generating whole genomes of modern and ancient organisms for the first time (Schmutz et al., 2004).[k]

In 2010, four years after the announcement, Pääbo and his team finally finished the Neanderthal Genome Project.[40] The project, conducted by over fifty scientists at a cost of approximately €5 million, successfully sequenced 4 billion base pairs of Neanderthal DNA.[35,41,42] Scientists, for the first time, declared they had data to answer their questions about Neanderthal evolutionary history, specifically their relationship to humans. However, it was not just the data that was important but the ability to

---

[i]The debate in evolutionary anthropology centred around the origins of human history with evolutionary anthropologists usually subscribing to one of two hypotheses; the Out of Africa Model or the Multiregional Continuity Model. The former proposes that humans originated in Africa and then migrated to other parts of the world, while the second suggests that pre-humans originated in Africa but then evolved into modern humans after they migrated out of the continent.[34]

[j]Svante Pääbo detailed the development of the Neanderthal Genome Project from a personal perspective.[35] Further, it is absolutely important to note that the feasibility of the project was in principal possible because of new technology, namely next generation sequencing technology, that had been developed by companies like 454 Life Science Corporation.

[k]By this time, there had also been a number of scientific, conceptual, and technological developments seeking to study the evolution and extinction of Neanderthals through ancient genetics.[33,37,38]

analyse it was a critical component of the project. The combination of this genomic data and statistical methods, developed by David Reich and his lab at Harvard University, that allowed them to detect signals of admixture (i.e. interbreeding) between early humans and Neanderthals.

With information from the genome, scientists inferred that early humans bred with their archaic ancestors — the Neanderthals — before going extinct 30,000 years ago. However, the evidence seemed to suggest that Neanderthals only interbred with a particular human population, those humans who had travelled out of Africa and into Europe. By comparing the Neanderthal genome with present-day human genomes, they determined that Neanderthals shared more similarities with present-day non-African populations than with present-day African populations. Neanderthal DNA existed in a small percentage (one to four percent) of a specific population (Eurasian population). *National Geographic* reported the results: 'The next time you're tempted to call someone a Neanderthal, you might want to take a look in the mirror. According to a new DNA study, most humans have a little Neanderthal in them — at least 1 to 4 percent of a person's genetic makeup'.[43] While Pääbo expected to engage the archaeological and anthropological community with these radical results, it seemed he had apparently not anticipated how the public would react. Their paper published in *Science*, for example, attracted attention from the creation-ist religious community in the United States who reinterpreted results as evidence for or against their own private projections about Neanderthals' relations to humans and creation. Women also wrote to Pääbo with their speculations that their own husbands were Neanderthals. *Playboy* even spotlighted the research in a four page spread titled 'Neanderthal Love: Would You Sleep with This Woman?'[35] (pp. 221–222).

The use of ancient DNA data to shed light on human evolutionary history did not stop there. Rather, scientists entered a race to sequence the first genomic data from ancient plants, animals, and diseases to human Paleo-Eskimos, Aboriginal Australians, and famous historical figures like King Richard III.[44-47] Scientists also sequenced the first genomic data from a Denisovan, an extinct hominin species whose identity as a distinct archaic human species was uniquely obtained exclusively from DNA, without a real fossil record and only from a small sliver of a finger bone.[48,49] Meanwhile, other research shed light on the behaviour of our early ancestors, including Mesolithic and Neolithic hunter–gatherers, while also exploring transformations in human cultural practices such as milk consumption which have directly impacted our evolution in terms of selection for lactase persistence.[50–52] Much work has also explored our interactions with animals through time by interrogating genetic signals for domestication in pigs, cattle, and dogs on large global and temporal scales.[53,54] According to some scientists and journalists, these works have led the way towards a revolution in our understanding of human origins, evolution, migrations, and our relatedness to each other.[55,56]

All the unexpected excitement about interbreeding with our archaic ancestors, the Neanderthals, has caused an onslaught of professional and popular interest, but what does the ancient and modern genomic evidence regarding the evolution, migration, and intermixing of past populations across the world and over time really tell us about the age-old question of the human species as it relates to the concept of race and relatedness? Reich's book, for example, spoke to this question. For Reich, the genetic evidence challenges our previous perceptions of race in two important ways. First, Reich rejected the old view of 'race' in terms of the opinion that there is such a thing as a 'pure race' (i.e. a holism as portrayed by the Nazi Party specifically and National Socialism generally)[5] (p. 267). *The Sunday Times*, for example, put the point this way: 'There's no such thing as racial purity — and now there's the DNA to prove it'.[57] *The Guardian* also reported on this result, explaining: 'The overriding lesson ancient DNA teaches is that the population in any one place has changed dramatically many times since the great human post-ice age expansion, and that recognition of the essentially mongrel nature of humanity should override any notion of some mystical, longstanding connection between people and place. We are all, to use Theresa May's derisive label, "citizens of nowhere"'.[58] In other words, ancient DNA data is helping to paint a picture of human complexity and unity — one people, one world, with lots of mixing over tens to hundreds to thousands of years.[i] In some ultra-holistic way, the ancient and modern genetic evidence seem to suggest the unity of humanity over time.

At the same time, however, Reich rejected the prevailing academic perspective that human populations are *so* closely related that there can be *no* significant biological differences between them. Since World War II, this was the academic orthodoxy — a conclusion reached through research by both anthropologists and geneticists that argued there were no significant differences between human populations to justify the concept of 'biological race'[5] (p. 250). Ancient DNA, as far as Reich and some other scientists were concerned, actually offers evidence for the fact that there *are* important genetic differences among human populations that *should be studied* because they provide insight into human health (i.e. risk factors for some diseases). Reich, for example, noted that prostate cancer occurs at a much higher rate in African American men than in European American men, and according to research results, genetic factors on specific regions in the genome were responsible for this increased propensity to disease. By comparing the genomes of more than 1,500 African American men, scientists found evidence of a connection between African ancestry and risk factors for prostate cancer. This was an important difference between human populations that they felt should not simply be ignored.

In light of this, geneticists find themselves between a rock and a hard place, as they attempt to explain what appears to the simultaneous nuances of genetic similarities and differences across human populations. Here, Reich suggested a new way of understanding the differences between human

---

[i]In 2014, *The New York Times* journalist Nicolas Wade published a highly controversial book, *A Troublesome Inheritance: Genes, Race, and Human History*, in which he argued for a genetic basis for the differences between human populations.

populations and reconciling the scientific data with our societal concerns tied to outdated opinions of race. He suggests that the 'genomic revolution' can help provide a 'rational framework' in which both professionals and the public can recognise the differences and similarities among past and present human populations without the baggage of history.

Conclusion

All this talk of the genomic revolution seems to suggest a much more holistic perception of human evolutionary history in terms of the multiple mixings of people across space and over time. However, it is striking that this new kind of holism is based on arguably one of the most reductionist sort of sciences — namely the science and technology of ancient DNA research where nearly everything comes down to a letter in a sequence. Moreover, it is striking that some scientists in this new field exhibit a tendency to embrace, intentionally or unintentionally, a reductionist mindset in regards to their choice of data (i.e. molecular data) as unsurpassed evidence for understanding human origins, evolution, and migrations.

According to archaeologists, geneticists sometimes embrace their evidence to the exclusion of other forms of evidence or interpretations of evidence like that from the fields or archaeology or linguistics. Although archaeologists are reacting to, and sometimes resisting, what some might see as a gung-ho attitude of acceptance towards ancient DNA data and just how to interpret it as it applies to questions regarding the human past, as well as what it might mean for society today. Sure enough, archaeologists and geneticists have teamed up together in their efforts to study human history, there is a growing group of archaeologists inclined to push back against the way in which ancient DNA data is being used to define human population structures and movements in the past.[59] Some recent editorial reviews in *Nature* have highlighted this angst among archaeologists.[60] Ewen Callaway, for example, reported on the reaction: 'Some archaeologists, however, worry that the molecular approach has robbed the field of nuance. They are concerned by sweeping DNA studies that they say make unwarranted, and even dangerous, assumptions about links between biology and culture'. According to Callaway, 'They give the impression that they've sorted it out', says Marc Vander Linden, an archaeologist at the University of Cambridge, UK. 'That's a little bit irritating'[60] (p. 574). These sentiments are far from a lone case of data envy. For archaeologists, this reductionist outlook has important implications for the public understanding of science.

Among the press and public, as well as within the scientific sphere, DNA has been repeatedly referred to as the 'book of life' or the 'code of life'. Scholarship has demonstrated how these metaphors have shaped professional and popular discourse around genetic and genomic research, colouring the public perception of DNA as a decisive and deterministic player in the making of life.[6,61] This reductionist view of life has become public discourse, especially when it comes to public understanding of human biology and human evolution. These metaphors suggest simplicity and lead the public to assume that life can be explained by a 'code' and furthermore, that scientists have the ability to decipher that code — whatever it may be. However, scholarship has also discussed the severe shortcomings of this rhetoric, particularly the injustice it does to the intricate interactions between molecules and the external environment, and at times the insensitive implications it can have for societal concerns more broadly.[6,61] In other words, DNA tells us a lot about what it means to be human, but it does not and cannot explain absolutely everything.

This sort of reductionist reliance on the power of genetic and genomic information to tell us the answers to our questions about 'who we are' and 'how we got here' is undoubtedly laden with historical and social baggage. One review remarked on Reich's goal to build an 'American-style genomics factory' for the study of ancient DNA research. For Maria Avila Arcos, this had potential problems: 'When one considers the social and historical context of the human populations that will be studied — many of which have been historically marginalised, colonised, and exploited — this statement becomes problematic. Such intentions could easily be perceived as a continuation of exploitation or biocolonialism'.[62] Avila Arcos took note of a point Reich made that unfortunately likens his work to problematic practices of former centuries: 'An unfortunate analogy further highlights the problem. We are ... like explorers in the late eighteenth century, sailing to every corner of the globe', he writes. During the era to which Reich refers, European adventurers indeed collected samples from around the world, but these specimens were usually taken without the consent of, or regard for, the communities to whom they rightfully belonged'.[62]

   To be clear, our mentioning of this criticism does not mean we are attacking Reich's position in particular or the science of ancient DNA research more generally as it applies to the study of human evolution. Rather, this comment and the controversy it conveys serves as a not-so-gentle reminder of the first point in this paper — that the science of human evolution is far from a private affair and that it comes with a hefty load of baggage. Indeed, Reich was more than aware of the history that haunts the science of human genetics, particularly as he addressed the well-known and widely condemned racist viewpoints of James Watson, one of the scientists responsible for determining the structure of DNA. With Watson close in mind, Reich 'shudder[s] to think of scientists who have preceded him and who have held scientifically and morally wrong views about population differences across the human species'[5] (p. 264). The genomic revolution, made possible by the search for DNA from ancient and extinct organisms a mere three decades ago, is coming up with exciting and unprecedented findings about human evolutionary history. At the same time, however, the scientists behind the science must understand that the data generated and interpreted in the lab has very real societal consequences outside the lab. The study of human evolution is a public one, and any scientific discussion necessarily includes, and also is influenced by, larger questions in the history and philosophy of science, including those that deal with the big question of holism, reductionism, and its relationship to the evolutionary history of the human species.[m]

References

1. Darwin E. (1825) *The Temple of Nature or, the Origin of Society: A Poem, with Philosophical Notes. With Plates, Including a Portrait.* Jones, London.
2. Darwin C. (1859). *On the Origin of Species by Means of Natural Selection, or the Preservation of Favoured Races in the Struggle for Life.* John Murray, London.
3. Dobson E. (2010). *Darwin's Circle: How His Scientific Collaborators and Critics Reacted to the Theory of Natural Selection.* North Carolina State University.
4. Barrett PH, Gautrey PJ, Herbert S, et al. (eds.). (1987) *Charles Darwin's Notebooks,* pp. 1836–1844. Cornell University Press: Ithaca, N.Y.
5. Reich D. (2018) *Who We Are and How We Got Here: Ancient DNA and the New Science of the Human Past.* Pantheon Books, New York.
6. Nelkin D and Lindee S. (1995) *The DNA Mystique: The Gene as a Cultural Icon.* Freeman, New York.
7. Smocovitis VB. (1996). *Unifying Biology: The Evolutionary Synthesis and Evolutionary Biology.* Princeton University Press, New Haven.
8. Dietrich MR. (1998) Paradox and persuasion: Negotiating the place of molecular evolution within evolutionary biology. *J Hist Biol* **31**(1): 85–111.
9. Hagen JB. (1999) Naturalists, molecular biologists, and the challenges of molecular evolution. *J Hist Biol* **32**(2): 321–341.
10. Lynch M, Cole SA, McNally R and Jordan K. (2008) *Truth Machine: The Contentious History of DNA Fingerprinting.* University of Chicago Press, Chicago.
11. Suárez-Díaz E and Anaya-Muñoz VH. (2008) History, objectivity, and the construction of molecular phylogenies. *Stud Hist Philos Biol Biomed Sci* **39**(4): 451–468.

---

[m]In this paper, we have focused on showing that, when it comes to human evolution, there is no such thing as good/bad holism/reductionism. Both approaches are needed, and care is also needed to see that they are not perverted for wrong ends. We are much aware that we have barely started on the project. There are, for instance, still contentious issues about whether there is any place for intraspecific grouping. Given its vile history, one's inclination is to stay far away from such thinking. Then one learns that African American men are twice as likely to get prostate cancer as white men. Is one to ignore this fact in the cause of political correctness? We ask this question. We do not answer it.

12. Creager ANH. (2009) Phosphorus-32 in the Phage group: Radioisotopes as historical tracers of molecular biology. *Stud Hist Philos Biol Biomed Sci* **40**(1): 29–42.

13. Strasser BJ. (2012) Data-driven sciences: From wonder cabinets to electronic databases. *Stud Hist Philos Biol Biomed Sci* **43**: 85–87.

14. García-Sancho M. (2012) *Biology, Computing and the History of Molecular Sequencing: From Proteins to DNA*, pp. 1945–2000. Palgrave Macmillan, London.

15. Rasmussen N. (1997) *Picture Control: The Electron Microscope and the Transformation of Biology in America*, pp. 1940–1960. Stanford University Press, Redwood City.

16. Rasmussen N. (2014). *Gene Jockeys: Life Science and the Rise of Biotech Enterprise*. Johns Hopkins University Press, Baltimore.

17. Yi D. (2015) *The Recombinant University: Genetic Engineering and the Emergence of Stanford Biotechnology*. University of Chicago Press, Chicago.

18. Sommer M. (2016) *History Within: The Science, Culture, and Politics of Bones, Organisms, and Molecules*. University of Chicago Press: Chicago.

19. Stevens H. (2016) *Biotechnology and Society: An Introduction*. University of Chicago Press, Chicago.

20. Jones ED. (2018) Ancient DNA: A history of the science before Jurassic Park. *Stud Hist Philos Biol Biomed Sci* **68**: 1–14.

21. Jones ED. (2018) Ancient genetics to ancient genomics: Celebrity and credibility in data-driven practice. *Biol Philos* **34**(27): 1–35.

22. Bösl E. (2017) *Doing Ancient DNA: Zur Wissenschaftsgeschichte Der ADNA-Forschung*. Verlag, Bielefeld.

23. Bösl E. (2017) Zur Wissenschaftsgeschichte Der ADNA-Forschung. *NTM Zeitschrift Für Geschichte Der Wissenschaften, Technik Und Medizin* **25**(1): 99–142.

24. Jones M. (2016) *Unlocking the Past: How Archaeologists Are Rewriting Human History with Ancient DNA*. Arcade Publishing, New York.

25. Shapiro B and Michael H (eds.). (2012). *Ancient DNA: Methods and Protocols*. Springer, New York.

26. Margulies M, Egholm M, Altman WE, *et al.* (2005) Genome sequencing in microfabricated high-density picolitre reactors. *Nature* **437**: 376–380.

27. Knapp M and Hofreiter M. (2010). Next generation sequencing of ancient DNA: Requirements, strategies and perspectives. *Genes* **1**(2): 227–243.

28. Cano RJ, Poinar HN and Poinar GO Jr. (1992) Isolation and partial characterisation of DNA from the bee Proplebeia dominicana (Apidae: Hymenoptera) in 25–40 million year old amber. *Med Sci Res* **20**(7): 249–251.

29. Cano RJ, Poinar HN, Pieniazek NJ, *et al.* (1993) Amplification and sequencing of DNA from a 120–135-million-year-old weevil. *Nature* **363**(6429): 536–538.
30. DeSalle R, Gatesy J, Wheeler W and Grimaldi D. (1992) DNA sequences from a fossil termite in oligo-Miocene amber and their phylogenetic implications. *Science* **257**(5078): 1933–1936.
31. Golenberg EM, Giannasi DE, Clegg MT, *et al.* (1990) Chloroplast DNA sequence from a Miocene Magnolia species. *Nature* **344**: 656–658.
32. Woodward SR, Weyand NJ and Bunnell M. (1994) DNA sequence from Cretaceous period bone fragments. *Science* **266**(5188): 1229–1232.
33. Krings M, Stone A, Schmitz RW, *et al.* (1997) Neandertal DNA sequences and the origin of modern humans. *Cell* **90**: 19–30.
34. Stringer C. (2012) *Lone Survivors: How We Came to Be the Only Humans on Earth.* St. Martin's Press, New York.
35. Pääbo S. (2014) *Neanderthal Man: In Search of Lost Genomes.* Basic Books, New York.
36. Green RE, Krause J, Ptak SE, *et al.* (2006) Analysis of one million base pairs of Neanderthal DNA. *Nature* **444**(7117): 330–336.
37. Ovchinnikov IV, Götherström A, Romanova GP, *et al.* (2000) Molecular analysis of Neanderthal DNA from the northern Caucasus. *Nature* **404**(6777): 490–493.
38. Krause J, Lalueza-Fox C, Orlando L, *et al.* (2007) The derived FOXP2 variant of modern humans was shared with Neandertals. *Curr Biol* **17**(21): 1908–1912.
39. Hofreiter M. (2008) Palaeogenomics. *C R Palevol* **7**: 113–124.
40. Green RE, Krause J, Briggs AW, *et al.* (2010) A draft sequence of the Neandertal genome. *Science* **328**(5979): 710–722.
41. *Max Planck Society.* (2010) The Neandertal in Us. https://www.mpg.de/617258/pressRelease20100430.
42. Callaway E. (2010). Neanderthal genome reveals interbreeding with humans. *New Scientist*, May 6.
43. Than K. (2010) Neanderthals, Humans Interbred — First Solid DNA Evidence. *National Geographic,* May. http://news.nationalgeographic.com/news/2010/05/100506-science-neanderthals-humans-mated-interbred-dna-gene/.
44. Gilbert MTP, Drautz DI, Lesk AM, *et al.* (2008) Intraspecific phylogenetic analysis of Siberian woolly mammoths using complete mitochondrial genomes. *Proc Natl Acad Sci USA* **105**(24): 8327–8332.
45. Gilbert MTP, Kivisild T, Grønnow B, *et al.* (2008). Paleo-Eskimo MtDNA genome reveals matrilineal discontinuity in Greenland. *Science* **320**(5884): 1787–1789.
46. King TE, Fortes GG, Balaresque P, *et al.* (2014) Identification of the remains of King Richard III. *Nat Commun* **5**: 5631. doi:10.1038/ncomms6631.

47. Willerslev E, Davison J, Moora M, *et al.* (2014) Fifty thousand years of arctic vegetation and megafaunal diet. *Nature* **506**(7486): 47–51.

48. Gokhman D, Lavi E, Prufer K, *et al.* (2014). Reconstructing the DNA methylation maps of the Neandertal and the Denisovan. *Science* **344**(6183): 523–527. doi:10.1126/science.1250368.

49. Reich D, Green RE, Kircher M, *et al.* (2010) Genetic history of an archaic hominin group from Denisova Cave in Siberia. *Nature* **468**(7327): 1053–1060. doi:10.1038/nature09710.

50. Izagirre N and de la Rúa C. (1999) An MtDNA analysis in ancient Basque populations: Implications for haplogroup V as a marker for a major Paleolithic expansion from southwestern Europe. *Am J Hum Genet* **65**(1): 199–207. doi:10.1086/302442.

51. Jones BL, Oljira T, Liebert A, *et al.* (2015) Diversity of lactase persistence in African milk drinkers. *Hum Genet* **134**(8): 917–925. doi:10.1007/s00439-015-1573-2.

52. Malmström H, Linderholm A, Skoglund P, *et al.* (2015) Ancient mitochondrial DNA from the northern fringe of the Neolithic farming expansion in Europe sheds light on the dispersion process. *Philos T R Soc B* **370**(1660): 1–10. doi:10.1098/rstb.2013.0373.

53. Leonard JA, Wayne RK, Wheeler J, *et al.* (2002) Ancient DNA evidence for Old World origin of New World dogs. *Science* **298**(5598): 1613–1616.

54. Bollongino R, Edwards CJ, Alt KW, *et al.* (2006) Early history of European domestic cattle as revealed by ancient DNA. *Biol Lett* **2**(1): 155–159.

55. Stoneking M and Krause J. (2011) Learning about human population history from ancient and modern genomes. *Nat Rev Genet* **12**(9): 603–614. doi:10.1038/nrg3029.

56. Veeramah KR and Hammer MF. (2014) The impact of whole-genome sequencing on the reconstruction of human population history. *Nat Rev Genet* **15**: 149–162.

57. Appleyard B. (2018) Book review: Who we are and how we got here: Ancient DNA and the new science of the human past by David Reich. *The Sunday Times*, April 1.

58. Forbes P. (2018) Who we are and how we got here by David Reich review — new findings from ancient DNA. *The Guardian*, March 29. https://www.theguardian.com/books/2018/mar/29/who-we-are-how-got-here-david-reich-ancient-dna-review.

59. Heyd V. (2017) Kossinna's Smile. *Antiquity* **91**(356): 348–359.

60. Nature editorial. (2018) On the use and abuse of ancient DNA. *Nature* **555**: 559.

61. Anker S and Nelkin D. (2003) *The Molecular Gaze: Art in the Genetic Age.* Cold Spring Harbor Laboratory Press, Cold Spring Harbor.

62. Avila Arcos MC. (2018) Troubling traces of biocolonialism undermine an otherwise eloquent synthesis of ancient genome research. Book review in *Science.* https://blogs.sciencemag.org/books/2018/04/17/who-we-are-and-how-we-got-here/

7 'Am I My Brain?' Neuro-Centrism and the Law

Harris Wiseman*

Introduction

Understanding the brain is an important part of gaining a scientific under-standing of who we are. Without question, our personalities, memories, and identities are mediated by our brains. But is that *all* we are? Can identity be understood *only* by looking at the brain? There has filtered through to popular consciousness a basic acceptance of this brain-centred approach to thinking about identity. Indeed, a brain-centric approach is too often deemed 'the real truth' about who we are. The idea that 'we are our brains', and that every other part of who we are is essentially dispensable, or replaceable, is commonly expressed in popular discourse and neuro-philosophy. Yet, nothing could be further from the truth.

In this chapter, I will consider the extent to which the brain, though essential for a scientific understanding of identity, is not sufficient for under-standing identity either. Then, I will consider how legal systems around the world are responding to this neuro-centric view of identity, given the purported implications such neuroscience arguably portends.

---

*Research Fellow at Campion Hall, University of Oxford, Brewer St., Oxford 0X1 1QS, UK.

'I Am (Not) My Brain!'

Let us take a look at some of the reasons why having a scientific under-standing of the brain is so important for understanding identity. After all, if my brain is damaged, do I not become a different person? If my brain activity is heavily dampened, do I not cease to be conscious? If my brain is subjected to intoxicants, do I not act like a completely different per-son? Must it not be the brain, then, that is in charge? There are clear and incontrovertible reasons for thinking that the brain is one crucial locus of our identity. Let us take some of the more striking examples of changes to the organic matter of the brain, directly followed by strong changes in the identity of the person in question.

Gage was No Longer Gage

Many are familiar with the case of Phineas Gage, a railroad worker, given prominence in the work of Antonio Damasio, who wrote of the effects on Gage's personality after surviving a giant tamping iron being fired through his head:

> Gage's disposition, his likes and dislikes, his dreams and aspira-tions are all to change. Gage's body may be alive and well, but there is a new spirit animating it.
>
> GAGE WAS NO LONGER GAGE
>
> … He walked firmly, used his hands with dexterity, and had no noticeable difficulty with speech or language. And yet, … He was now fitful, irreverent, indulging at times in the gross-est profanity which was not previously his custom, … So radical was the change in him that friends and acquaintances could hardly recognize the man. … So different a man was he that his employers would not take him back when he returned to work … The problem was not lack of physical ability or skill; it was his new character.[1] (pp. 8–9, emphasis added)

In short, we find here a man once having had a pleasant and genteel character, then undergoing a distinct change in character after having his brain damaged.

Links between brain changes and changes in character have long been observed, not just in the scientific literature,[a] but in more ancient sources relating to practices of 'trepanning' (drilling holes in the skull to relieve pressure on the brain, performed potentially in prehistoric times, but certainly from Roman times onwards), as well as being enshrined in the medieval theory of 'humours' (Don Quixote's madness was comically attributed to him by the author as a consequence of the drying up of the liquids in his brain, caused by reading too many romances). In other words, linking the brain with identity and personality is nothing new; and as Hippocrates suggested in approximately 430 BC 'Men ought to know that from nothing else but thence [from the brain] come joys, delights, laughter and sports, and sorrows, griefs, despondency, and lamentations'.[5]

Looking at the more recent science, one relatively consistent set of findings seems to surround the following types of personality changes resulting from neurological damage to the frontal lobes: 'disturbed social behaviour, executive/decision-making deficits, diminished motivation/hypo-emotionality, irascibility, and distress'[4] (p. 833). Even so, physiological damage to the brain can result in bizarre changes to a person's character which are highly unpredictable, and, there remains a huge amount of mystery as to the precise mechanisms in play.

Deep Brain Stimulation and Inadvertent Effects of Neurosurgery

While most of the studies referred to above seem to focus on the negative impact of neurological changes on personality, there is more recent work

---

[a] Studies traverse a wide timespan, from at least 1952 to 2011.[2–4]

on positive changes to identity after neurological alteration. Marcie King *et al.* have investigated positive personality changes inadvertently occurring after surgery for other neurological conditions (e.g. to repair damage done by a stroke, or benign tumour resection). King *et al.* write:

> Research on changes in personality and behavior following brain damage has focused largely on negative outcomes, such as increased irritability, moodiness, and social inappropriateness. However, clinical observations suggest that some patients may actually show positive personality and behavioral changes following a neurological event. … Lesion analyses indicated that positive changes in personality and behavior were most consistently related to damage to the bilateral frontal polar regions and the right anterior dorsolateral prefrontal region. These findings support the conclusion that … such changes have systematic neuroanatomical correlates.[6]

These findings are expressed yet more vividly by Christian Jarrett, who reports on one particular case: '"Patient 3534", a woman who had a brain tumour removed at the age of 70, leaving damage to the front of both sides of her brain. According to her husband, who'd known her for 58 years, before her surgery she had a "stern" personality, was highly irritable and grumpy. After the brain surgery, he said that she was "happier, more outgoing, and more talkative than ever before."[7] Taken on its own, such a finding might not seem remarkable, but against the body of evidence that is being accrued for sudden positive and negative changes produced by surgery to the brain, it becomes increasingly transparent that the physical structure of the brain is intimately related to personal identity.

Such studies raise powerful questions. Could a sufficiently fine-grained understanding of the relationship between identity and the physical substance of the brain be elaborated to alter identity *at will*? Such questions of 'psychosurgery' are not new. Electroconvulsive therapy and transorbital lobotomisation were applied in the 1960s and 1970s under medical and psychiatric rubrics for treating aggressive and psychotic behaviour. And today, Deep Brain Stimulation (DBS), or the less invasive neuro-therapies (for example, transcranial direct current stimulation [tDCS] or transcranial magnetic stimulation [TMS]), are used for the treatment of major depression and chronic anxiety.

DBS is a highly invasive treatment that has been used for many decades, originally to help manage Parkinson's disease. An electrode is implanted deep into the brain tissue, the region of the brain responsible for dopamine and serotonin release, and for regulating motivation and mood (the nucleus accumbens). Taylor and Weatherspoon write: 'Although doctors aren't exactly sure why the pulses help the brain reset, the treatment appears to improve mood and give the person an overall sense of calm'.[8] DBS is certainly taken as a more extreme therapy because of medical dangers (infection, allergic responses, brain haemorrhage, the need for maintenance surgery, etc.). However, documented accounts in the medical literature of changes to mood which are powerful, instant, and enduring, are very striking, though such changes are not always for the better (Phem et al.[9] note a significant increase in impulsivity in Parkinson's patients receiving such therapy, which can be a good or bad thing depending on one's context). Again, this raises powerful questions for persons confronting questions about their identity, questions about how durable our identities are, and how susceptible to medical intervention our very selves might be.

Dementia and Concussive Trauma and Contact Sports in Schools

While Damasio has been accused of misrepresenting Gage's story above, neglecting the fact that, over time, Gage was gradually rehabilitated,[10] the basic idea that alteration of the physical substance of the brain can have profound effects on a person's identity has been more than amply validated. Dementia is a very clear case in which deterioration of a person's brain has profound effects on personal identity. According to the Alzheimer's Association website: 'Dementia is a general term for loss of memory and other mental abilities severe enough to interfere with daily life. It is caused by physical changes in the brain'. With Alzheimer's (approximately 60–80% of dementia cases are of this order), imaging allows one to see certain 'hallmark abnormalities', such as 'deposits of the protein fragment beta-amyloid (plaques) and twisted strands of the protein tau (tangles) as well as evidence of nerve cell damage and death in the brain'.[11] Dementia shows how a loss of the substance of the brain, and a loss in its connectivity, does indeed result in cognitive impairments, as well as resulting in profound shifts in behaviour, personality and identity wherein the subject is described as, quite literally, no longer being the same person.

Confronting all the above evidence, profound ethical and legally-pressing issues are quickly raised to the surface. One of the instances in which such neuroscience has extremely significant, and urgent, legal ramifications relates to concussive trauma produced by school sports. Newer neuroimaging techniques allow a much improved ability to see the effects of damage to the brain. Like dementia, of which Chronic Traumatic Encephalopathy (CTE) is considered a form, CTE is a degenerative condition affecting persons who have taken repeated concussive impacts to the head, those involved in rough contact sports, military combat, and long-term victims of domestic violence.[12] CTE was first described in American football players as a progressive neurodegenerative syndrome leading to neuronal loss coupled with protein and plaque deposits in the brain.[13] Symptoms of the brain damage created in this way include aggression, suicidal tendencies, mood changes, behavioural problems, difficulty thinking, and memory issues.[14] Studies performed on (English) football players have suggested that there are immediate effects on cognitive capacity coming from heading a ball.[15] And, as dementia.org reports: 'Repetitive concussions and head injuries are thought to scar brain tissue, damage the cerebellum and cause long-term damage to cerebral blood vessels'.

The social and legal implications regarding consent are controversial here. What of children playing rugby or football at school? School sports are precious and important, culturally and in terms of encouraging the physical health of young persons, but at what cost? No child is able to give proper consent to such significant potential harm in later life. Compelling school children to play rugby, for example (which results in considerably more concussive events than football), might very well be considered a form of child abuse, perhaps even criminally negligent. But, again, such a suggestion is highly controversial.

Neuroscience and advances in brain imaging techniques make the level of damage being done to children's brains by participating in such sports increasingly clear. While longitudinal studies specifically relating damage caused by such contact sports on children's brains are still underway, the terrible long-term effects of brain damage on child development is known from other sources. Michael Rutter suggests: 'Brain injury causes a markedly increased risk in both intellectual impairment and psychiatric disorder' (but he does note that 'there are few psychological sequelae that are specific to brain damage'[16] [Rutter p. 1533]). Such damage done to a child's still developing brain is only amplified over the course of the lifespan. And, one might argue that subjecting a child's brain to such injury, longitudinal studies notwithstanding, is foolhardy at best. Certainly, studies showing the damage done to the brains of NFL players in the USA after scant years of play is visibly similar to the neuroimagery gathered from cases of dementia as outlined above, with tragic effects on the person's suffering such trauma, marked by the suicide of prominent ex-NFL players.[17]

Given that CTE has a latency period of up to a decade,[18] and the terrible effects of the disease, the stakes could not be higher. The issue must be addressed at the legal and social level. Issues of consent and informed choice at the adult level, engaging in sports such as boxing, or joining the military, may be a little more straightforward. However, neuroscience must be a crucial player in helping to guide decision-making for vulnerable persons involved in sports in order to prevent potentially lifelong, devastating consequences.

A Broader Perspective: Boundaries of Brain and Identity

Evidently, the neuroscience of identity has a great deal to offer, and certainly, the brain is one essential component in the system of factors which influence our identities. At the same time, the assumed boundaries of identity, which focus much too narrowly on the grey matter, must be expanded. This can be done by looking at the brain *in relationship* with many factors, at least:

(a) The brain's surrounding biology (i.e. considering embodiment)
(b) The larger social factors in the world that we all inhabit (i.e. considering enworldment)
(c) The wide range of human self-reflection which has shaped our self-concepts through the span of human civilisation

There are numerous empirical factors, not just from neuroscience, but from social and psychological sciences, that need to be considered when proposing a scientific account of identity, none of which are necessarily centred on the brain *per se*. For example, the social sciences are much more concerned with investigating identity in terms of environments, the roles persons play, the groups we are a part of, the contexts across which our identities seem to shift and change (right now I am a teacher, later tonight I will be a parent — my behaviour, comportment and sense of self all change accordingly, often radically); as well as the mentors and exemplars who model for us how to behave. All of these are completely non-neuro-centric foci. They will indeed be mediated by the brain, and they may very well have neurological correlates, but the point is that they need to be thought about primarily on their own level, and are usually not helpfully reduced to neurological language (just as it is not very helpful, when describing a recent holiday to a friend, to reduce talk about the trip to physics-based terms such as the nitrogen content of the air you were breathing while away — such terms, while technically accurate, do not convey what is really salient about the matter). To understand how my circumstances affect a shift in my identity, those circumstances have to be looked at on their own terms. Such factors tell us many essential things about how persons come to take on given identities, and how we sustain our different identities over the course of our lives.[19,20] Such factors need to be considered when discussing something as complex as human identity, and this is why a conversation between neuroscience and non-neurologically focused domains is so important.

Neuroscience is necessary, but insufficient, as a means for capturing identity in scientific terms. While that proposition may seem to go against popular opinion, it is helpful to be reminded that many practising neuroscientists are very hospitable to such multi-scientific engagement — particularly in the present data-driven milieu, which is all for combining models and data from as many scientific disciplines as possible.[21] Indeed, one might say that neuroscience is constantly pointing beyond itself. At the very least, practising neuroscientists are eager to engage with genetic and epigenetic data streams where available. Moreover, since identity is necessarily interpersonal, the development of social neuroscience,[b] and a philosophically-informed neuroscience which is welcoming of more broadly socio-historical contextual factors, are now coming to be understood as being crucial for taking our understanding forwards.[22] In reality, the suggestion that brain-talk is the only scientific way of talking about identity is certainly not held by neuroscientists themselves, who willingly rely on insights coming from other scientific domains to inform and guide their hypotheses and experimentation.[c]

---

[b]This social neuroscience perspective is extremely significant because it suggests that, even taking identity to be just 'in the brain', my identity cannot be understood only in reference to *my* brain, because my brain is mirroring and synchronising with the activities of the brain of another person.

[c]All this should be made clear before one even gets to making the point that identity can be talked about validly in completely non-scientific terms (poetically, existentially or experientially, and so on), and indeed must be talked about non-scientifically if one is to have a rich and broad *human* understanding of who we are.

Yet, even accepting that brain-talk is not the only valid way of approaching identity, there are those sympathetic towards the view that the brain is still the *centrepiece* of the matter. Instead, if one takes as one's starting point the idea that one's world and one's relationships are every bit as influential as one's brain when it comes to sustaining one's identity, then a different picture emerges. Taking the perspective that we are persons in a world, a world of others, having bodies and brains and minds, having histories and cultures, knowing that all of this has some ongoing, dynamic impact on our identities, then it becomes harder to make sense of this popular view that 'I am my brain'. The truth is that there is more to identity than just the focusing on the brain will allow us to grasp. The brain is not a completely self-enclosed entity whose functioning can be understood solely by reference to itself. Again, this is a point which most neuroscientists themselves find uncontroversial.

Simply put, there is world of difference between making the brain the centre of study because that is one's area of interest and expertise; and making the brain the centre of the conversation because one believes that it is simply the centre of everything, around which everything else naturally revolves, and in terms of which everything else ought to be described. The former is necessary and valuable, the latter holds not a shred of truth.

Let us explore the limits of thinking of identity in neuro-centric terms. A colourful illustration of how neuro-centrism can run amok can be found in cryogenics. Neuroscientist Sergio Canavero (who claims that human head transplantation will occur before 2021[23]), has already been given consent by living patients to take hold of their brains. Many of these volunteers are dying of incurable diseases or have been paralysed through injury. The promise is that these brains will be transplanted into a completely new body, letting the person live on liberated from their current bodily infirmities. There could not be a more decisive image of the neuroreduction of identity. The entire body is being considered as completely disposable, irrelevant to one's identity which is implicitly assumed to be stored wholly in the brain.

But there are many important problems here, for the brain is not a wholly discrete organ. By way of contrast, one can readily cut out a kidney or a liver for transplantation — one knows the boundaries of these organs well enough. But, the brain, which is itself modular, a closely interwoven set of interconnected substructures, is intimately tied into the central nervous system. The nerves, ordered by the spinal column, go directly into the brain and form an intimate part of the brain. Precisely where does the brain end, and where does the central nervous system begin? Cryogenic practice has to grapple with this 'brain–body problem'. The assumption, in practice, is that the brain can be cut out, leaving one's identity wholly intact. But, where exactly is one to cut? The brain is indistinguishable from the rest of the central nervous system, and that itself is indistinguishable from the peripheral nervous system. The boundaries that one usually thinks of as separating the brain from the rest of the body are not necessarily so clear as one might think.

One example of how the brain might be said to exist beyond the grey matter (or, put better, that the brain functions beyond the boundary that popular culture and neuro-centrism has set up), is found in recent research on the relationship between the gut, its microbial contents, and brain function.[24] Leclercq et al. write:

> Gut bacteria strongly influence our metabolic, endocrine, immune, and both peripheral and central nervous systems. Microbiota do this directly and indirectly through their components, shed and secreted, ranging from fermented and digested dietary and host products to functionally active neurotransmitters including serotonin, dopamine, and γ-aminobutyric acid[24] (p. 204).

The so-called 'gut feeling' associated with the sense of intuition is more than just metaphorically reliant on activity in the gut. As microbiologist John Cryan writes:

> The concept of the gut influencing brain and behaviour has existed for almost two centuries. … despite this concept being

widely integrated into our everyday vernacular (gut feelings, gut instinct, gutted, gutsy, it takes guts, butterflies in one's tummy), it was not until the advent of brain imaging that neuroscientists really began to appreciate the influence of this axis on modulating brain function and maintaining homeostasis, especially during stressful situations.[25] (p. 201)

Indeed, as Siri Carpenter reports, the gut can be thought of as:

... the 'second brain,' it is the only organ to boast its own independent nervous system, an intricate network of 100 million neurons embedded in the gut wall. So sophisticated is this neural network that the gut continues to function even when the primary neural conduit between it and the brain, the vagus nerve, is severed. ... evidence has mounted from studies in rodents that the gut microbiome can influence neural development, brain chemistry and a wide range of behavioral phenomena, including emotional behavior, pain perception and how the stress system responds.[26] (p. 50)

Carpenter goes on to describe the gut's 'multifaceted ability to communicate with the brain' and notes that 'it's almost unthinkable that the gut is not playing a critical role in mind states'[26] (p. 50).[d] Likewise, Cryan notes that 'the ability to target the brain via the microbiome is viewed as a paradigm shift in neuroscience and psychiatry' (Cryan 2016, 202), even going so far as to speak of 'psychobiotics (bacteria with positive effects on mental health)'[26] (p. 50), putting focus on gut bacteria's potential to serve a protective function in developing adolescent brains, and having healing potential in treating PTSD. Perhaps this all sounds a little too far-fetched, but at the very least one can say that our mood, our development, and, over time, more enduring parts of our identity, are to some extent mediated

---

[d]This is not just an argument about whether the word 'brain' should be replaced by some phrase like 'nervous system' or 'gut' — that would be equally reductive, simply shifting one label to some other biological part. The point with embodiment is that the human body works as an integral whole, that identity is formed and sustained in the interaction of systems.

by the activities of our guts. So, one's identity is not just rooted in the grey matter of the brain, but directly rooted in the ongoing functioning of the gut, its symbiotic reliance on the microbial activity going on within it, and by extension related to a huge number of the biological activities the body is constantly performing and subject to. Identity and the brain both extend beyond the grey matter, the boundaries of which need to be redrawn by those relying on neuro-centric frameworks.

But, why stop there? What is a gut without a body to hold it together? And, for that matter, what is a body without a world to live in and respond to? Such a broader outlook is important for responding to the claims of strong neuroreductionists, like psychiatrist Mark Salter, who boldly made the following outlandish statement in a recent debate on neurology and free will: 'You see, Stalin's *central nervous system* killed millions of people. Shakespeare's *central nervous system* wrote great poetry'[27] (italics added). This sort of assertion, made in a public intellectual forum, shows just how far cultural neuro-centric assumptions have led astray the minds of otherwise seemingly intelligent persons. In reality, Stalin's central nervous system did not kill anyone, nor did Shakespeare's central nervous system write any poetry. One can rest assured that if a person's central nervous system is excised from his or her body and laid on a table, it will not be writing any poetry.

A central nervous system cannot write poetry, because without the connection to a body and a world in which others live, in relation to which there are things to write poetry about, a history, a language, a culture, there is nothing to talk about. There will be no writing poetry without any of these things, they are all essential. *So, why single out the brain?* The general retort to this might be something like: 'Because damage to Shakespeare's brain would have had more effect on his poetry than damage to any other of his organs'. Is that so? There are plenty of forms of damage, physical, emotional, relational, which can have a direct, enduring, sudden, and dramatic impact on one's identity. To lose an arm or a leg is indeed to lose a piece of oneself. One's life will never be the same again, and depending on the severity of the damage, one's identity may have to alter dramatically to accommodate that. One must reconfigure oneself anew, learn how to live one's life with an altered identity, as 'disabled', as a person no longer able to do the things one did before, a person seen differently and treated differently by other members of society. The impact to identity here is every bit as decisive as having a rail road spike through the brain was to Phineas Gage, yet these factors seem to be diminished, or ignored, in contrast to the common assumption, 'my brain is who I am, my brain is the house of "me"'.

Likewise, to lose a close family member is, perfectly literally, to lose a part of oneself also. Not only is identity not bounded by the brain, it is not even bounded by one's own body. We live in other people, and this extension of our identities is every bit as real as the neurological mediation of our identity, and every bit as prone to damage if extreme changes are introduced.

It needs to be emphasised that humans are meaning-making creatures, and that identity is also, to some considerable extent, bounded in terms of meaning. Above, we mentioned the retort that damage to Shakespeare's brain would have had more impact on his poetry than damage to his other organs. That was not the case with poet John Milton, author of Paradise Lost, who gradually became blind over the course of his life. Milton's case is very instructive here in showing how meaning-making can amplify the effects of the loss of a non-crucial biological faculty in an extreme way. Milton, a profoundly religious person, believed that the loss of his eyesight was an act of God, quite literally, a sacrificial exchange which gave him the authority to speak as God's witness. For Milton, Paradise Lost was a new gospel to be taken as seriously as the four found in the Bible, a new revelation from God. Milton's identity as a poet, which was the core of his self-defined identity, was exalted by the loss of his eyesight, reformed and radically reshaped thereby — likewise the effects on his ability to write poetry. Milton saw himself as having been given the highest and most authoritative sanction, no less than by God. Through meaning-making, the loss of Milton's eyesight had an unspeakably profound and enduring impact upon his identity and his ability to write poetry.[e]

This mindset that the brain is the core part of our body, the part that houses and determines our identities, ignores virtually everything important about our relationship to ourselves and to our world. Given the meaning-making creatures that we are, it needs to be understood that dramatic changes to any part of our world can come to produce at least as dramatic changes to our identity, which can be just as sudden and dramatic as any identity change noted in the neurological cases above. The brain is certainly *one* crucial locus of our identity — but it is not the only one, nor necessarily to be held as being more significant than any of the others. Again, the way we see identity comes down to our starting point. If one axiomatically assumes that the brain is the centre of the discussion, then one interprets all other factors in neurological terms. If, instead, one sees all these broader factors as being crucial and impactful (which they are), then there is no warrant for putting the brain at the centre of the discourse and comprehending everything in relation to its operations. At best, the grey matter contains the potential for identity, which can only be developed in relation to a world and to other people. It is the relationships between all the relevant factors, the ceaseless interactions between them that matter most of all. In this sense, my identity is not in my brain, nor it is in my body, nor is it in my world, nor in my circumstances. To reduce identity to any one of these factors would be to make the same mistake (as when one says: 'We are all just products of our environment' — a proposition which is just as reductive and erroneous). All these factors form the *theatre* in which our identities and relationships are developed and dynamically maintained in relation to other players we share the stage with, so to speak, all of whom

---

[e] This helps underline an important point that there are so many different ways of approaching identity: self-identity, perceived identity, narrative identity, to name but a few of the ways identity can be conceptualised. A full exploration is not possible here, but it is worth asking how many of these can be adequately captured in neurological terms.

have their own identities, all of whom 'play many parts'.[f] It is foolish to think of our personal identities without reference to this larger theatre of interactions. Viewed in such broader terms, the very notion that identity can be wholly reduced to one's own brain — that 'I am my brain' — ceases to make any sense at all.

Neurolaw: How Does Neuroscience Change the Way We Think about Criminal Responsibility?

These arguments regarding the complexities of identity, its relational and contextual character, and the inadequacies of reducing such matters to neuroscientific brain-talk, have profound implications for the enforcement of the law. What role can neuroscience play in helping to determine a person's guilt or innocence, their ability to stand trial? At present, legal systems around the world are wrestling with the new kinds of evidence that neuroscience is able to provide on identity and human agency. On the one hand, neuroscience is being used to inform certain kinds of criminal matters, and legitimately so. On the other hand, nothing really radical has changed, nor is it likely to, in the way criminal matters are settled in light of recent advances in neuroscience. Precisely because it is understood that the relationship between the brain and crime, or innocence, cannot be precisely fixed (because of its contextual and dynamic nature), there has been no neuro-revolution of the legal system, and contrary to popular neurophilosophical presentation of the supposedly illusory character of conscious choice, such a revolution does not seem to be fast approaching. There is some discussion in the popular press, stemming from popular neurophilosophers, regarding how criminals ought to be punished if free will is illusory. Joshua Greene, for example, presents the case for dispensing with retributive justice altogether (which would imply that we have free will, that criminals were free to have done other than they did), and favours a punitive system which takes deterrence as its ethical justification.[28]

---

[f]Of course, the famous image from Shakespeare is helpful for making this point!

Legislators have shown themselves to be anything but convinced by the evidence such neurophilosophers draw upon. In fact, a double-edged neuroscepticism (first, the extent to which neuroscience should be used to settle matters of law; and second, the extent to which present neuroscientific tools are even capable of settling matters of law), holds sway in the practice of the law around the world. While the popular press may be interested in sensationalist questions regarding neurodeterminism, the serious debate amongst jurisprudence scholars, legislators, and ethicists is much more concerned with where the lines are to be drawn between legitimate and illegitimate uses of current science;[9] and how the law is to keep pace with a field, nascent as forensic neuroscience is, whose forward development is far from being a straight line.

Mental Disorders and Diminished Responsibility: The 'Irresistible Impulse Defence'

The real questions neuroscience poses to jurisprudence overwhelmingly surround medically-related issues. There are cases where neural activity can be related to behavioural consequences. At present, by far the most significant set of questions which bring neural matters into contact with jurisprudence is related to issues of diminished responsibility. There are cases of neurological damage or disease in which neuroscience can be of clear assistance in evaluating 'the influence of a mental disorder on a defendant's decision-making process'.[30] Ongoing research is leading researchers 'to a greater comprehension of what had been termed *irresistible impulses* to commit crime and of the impact of brain damage, particularly evidence of brain lesions and frontal lobe damage on behaviour'[31] (p. 88 emphasis added). It is important to grasp that there are certain circumstances under which brain disorders, and certain mental health

---

[9]An example of where neuroimaging is influencing debates surrounding the legality of state practice can be found in the neuroscience of torture. Imaging techniques unambiguously show the neurological harms produced by torture.[29]

conditions, do provide a legitimate reason for accepting the diminished responsibility defence. Neuroscience can certainly be used to inform such cases. Working out which cases can be so informed is part of an ongoing process of developing a credible, evidentiary base for making such assessments. Current neuroscience is at a relatively early stage here, and this is represented in the justifiably cautious attitude taken towards neuroevidence in courts of law.

There are various impediments to effectively legislating for the validity of neuroevidence in determining diminished responsibility. The first problem is that neurological damage is very rarely of an all-or-nothing kind. When there are questions of disease or damage to the brain, the primary problem is determining the extent of the damage, and then mapping that onto a person's ability to understand, or resist, the behaviour that was perpetrated. Such determinations are, and will always be, more or less interpretive. As such, neuroimaging can only ever offer one vector of evidence to be considered amongst a larger body of contextual factors. It can settle very little in itself.

A wider evidentiary approach to settling criminal matters is important. This is precisely the approach that is currently taken in courts of law across the developed world. As Gkotsi and Gasser[32] (p. 25) point out: 'Neuroscientific data, no matter how accurate and reliable it may become, will only make sense in the quest for the assessment of criminal responsibility if they are contextualised and supplemented with data collected from other levels of analysis'. Neuroimaging and neuroscientific data do not always speak for themselves but need to be interpreted within the context of the crime in question. Gkotsi and Gasser continue: 'Despite the importance of the contribution of neuroscience to forensic assessments for a more refined understanding of the complex interaction between the brain, mental states and behaviour, the use of neuroscientific evidence in legal contexts will not dispense with the need to define the limits of responsibility and irresponsibility of the accused. It is a social, moral, political and, ultimately, legal question'[32] (p. 25).

One significant practical factor holding back the use of neuroscience in courts of law is that the science is fast-developing, and thus lines for determining the extent of diminished responsibility are still being hotly debated. Such lines are being drawn, redrawn and are continually being argued over. Issuing convictions or appeals on the bases of such blurred lines leaves court cases open for being overturned every time a new study is revealed, potentially contradicting the validity and weight of evidence previously given in court (an example of this will be given presently). This sort of issue is costly, both to the state, and with respect to the lives of persons whose futures are being determined on the basis of such evidence. Even restricting use of imaging techniques to extreme cases, controversies remain. For all the insights imaging provides, these techniques remain, for now at least, relatively crude.

Moreover, there are some neurological issues that cannot be dealt with using imaging data. For example, one complexity with neuroimaging comes with attempting to determine agency in mental health problems that have acute phases (e.g. schizophrenia), or mental conditions of short duration.[31] Herein, the person involved may not have comprehended what he or she was doing, and neuroimaging has significant limitations in such cases. Unlike with issues of enduring neurological damage, one cannot directly image temporary and historical changes in brain chemistry and function. Imaging can never conclusively indicate if a given person was in the grip of an acute phase reaction at a given moment in the past, though some indirect indications may be provided through neurological analysis.

While the law should certainly take account of credible scientific advances,[31] sometimes considerable quantities of time are required before one can judge how credible a given scientific technique is, or how useful in courts of law it might be. The general attitude of caution and neuroscepticism with respect to the law has (rightly) been heightened by significant failures in neurotechnology. One dramatic case of the utter failure of neurotech in determining truth and guilt, potentially having extreme and lifelong consequences, was the use of Brain Electrical Oscillations Signature

(BEOS) — so-called 'mindreading' neurotech — to secure the conviction of a woman for killing her husband in Indian courts in 2008. The BEOS technique concluded that the woman in question had been telling lies about her whereabouts, and used that determination to secure a conviction — a decision met with international consternation (pp. v–vii). It is interesting to note that matters of neurolaw form part of an international conversation, with scholars and ethicists keeping a close eye on the legal decisions and use of neuroscience in courts across the world. Legalists from six Western countries wrote to oppose the decision, and the conviction was eventually overturned.

Similarly, in the USA, mindreading companies (or 'brain fingerprinting' technology as it is otherwise called), such as No Lie MRI, and Cephos, have been employed to glean evidence for use in criminal cases. Even though such evidence was only used as part of a larger evidentiary picture, one conviction for a double murder was overturned using such technologies, a crime committed by Jimmy Ray Slaughter. Brain fingerprinting was then debunked, and Slaughter's appeal was overturned. He has since been put to death.[34] This is an example of the severity of the decisions that such neurotech and neuroevidence are being burdened with. The complete U-turn regarding the validity and reliability of certain kinds of neuro-based evidence only adds to the neuroscepticism, and to the general mood of caution with respect to the present use of such evidence in court.

Science does need to be taken seriously by the law, but in a balanced, discerning and context-sensitive manner. Brain fingerprinting technology is not admissible in courts in the UK, USA, Germany, or in India[33] (p. v). And, the practical reality of neuroevidence used in courts helps underline the very large gap between public presentation with its vivid neurosensationalism, on the one hand; and the actual practice of jurisprudence in the real world, on the other. While the public press remarks on such potential 'mindreading' tech in terms of the implications for privacy and human dignity, in the real world the questions are much more basic: to what extent does such tech work in the first place, can it provide reliable evidence in life-and-death cases, and what weight should be given to such evidence in relation to other evidentiary modes?

Finding a Middle Path in Neurolaw

While courts around the world seem to be moving at different paces, it is still easy to overstate the impact that neuroscience is currently having on them. Claydon writes: 'Neuroscientific advances are already informing court deliberations in England and Wales: assisting in considerations of guilt, fitness to plead and in sentencing'[31] (p. 88). Even so, not a huge amount has really changed when taking in the larger picture. At present, use of neuroscience as evidence in courts of law proceeds on a case-by-case basis.[28] According to Schleim[35] (p. 104): 'Evidence for an impending normative "neuro-revolution" is scarce'. While neuroscience is likely to gradually improve legal practice in the long run, Schleim continues: 'applying neuroscience methods about an individual's responsibility or dangerousness is premature at the present time and carries serious individual and societal risks'[35] (p. 104). Instead, advances in neuroscience must always be placed in a broader context that considers larger psycho-social factors.

As I have insisted throughout this chapter, our concepts and praxis in such matters go awry when they become neuro-*centric*. The problem is not just in making the brain the centrepiece of the conversation (when it is one crucial factor amongst many). The problem is the conceptual starting point, which conceives the grey matter as a separate entity existing in an isolated void, and as having all final causal responsibility attributed to its wiring and chemistry. As discussed above, neurological influences emerge in a context — the brain at once shapes, *and is shaped by*, its surrounding biology and environment. This process is dynamic and continuous. As Markowitsch and Staniloiu write:

> Despite persistence of terminologies such as 'genes of violence'
> or 'born to be criminal' in both scientific literature and media,
> a careful review of recent advances in neuroscience, genetics,

epigenetics and neuroimaging nowadays paints a more nuanced and complex picture of the current neuroscientific understanding of the underpinnings of individual violent behavior. Individual violent behavior can be viewed as a complex behavior that arises from a dynamic and likely time-sensitive interplay between genes and environment (including personal and cultural environment). Advances in the epigenetic field have taught us that genes (nature) and environment (nurture) cannot anymore be seen as separate, additive entities. Instead environmental factors can influence gene expression and brain development and synaptic plasticity in a time-dependent fashion.[36] (p. 1248)

Above, we concluded that the idea 'I am my brain' is misleading and false. This has profound implications for neurolaw. One's identity is not just in one's brain, and likewise, one's criminal culpability cannot be centred in the brain either. Neuroscience can be, and increasingly will come to be, a potent force in courts of law, offering valuable medically-related insights. These insights, used wisely, can indeed provide valid evidence to be considered regarding fitness to stand trial and culpability in cases of neurological damage. But there will always be limits to how far this mode of evidence can go. Except in cases of extreme neurological damage, neither identity nor criminal responsibility are adequately described in neurological terms. Though neuro-centric advocates may want to dispute the suggestion, the current position of caution and discernment towards neuroevidence in courts of law is, and will remain, the wisest policy.

References

1. Damasio A. (1994) *Descartes' Error: Emotion, Reason and the Human Brain*. Putnam Publishing, New York, NY.
2. Goldstein K. (1952) The effect of brain damage on the personality. *Psychiatry* **15**(3): 245–260.
3. Jennet B and Bond M. (1975) Assessment of outcome after severe brain damage: A practical scale. *Lancet* **305**(7905): 480–484.
4. Barrash J, Asp E, Markon K, *et al.* (2011) Dimensions of personality disturbance after focal brain damage: Investigation with the Iowa scales of personality change. *J Clin Exp Neuropsychol* **33**(8): 833–852.

5. Adams F. (trans.) (1972) *The Genuine Works of Hippocrates.* Robert E. Krueger Publishing Co, Huntington, N.Y. (http://www.humanistictexts.org/hippocrates.htm#Functions of the Brain) [18 October 2018].

6. King M, Manzel K, Bruss J and Tranel D. (2017) Neural correlates of improvements in personality and behavior following a neurological event. *Neuropsychologia* **21**Nov pii: S0028–3932(17)30445–1. doi: 10.1016/j.neuropsychologia.2017.11.023.

7. Jarrett C. (2018) When personality changes from bad to good. *BBC.* http://www.bbc.com/future/story/20180108-when-personality-changes-from-bad-to-good 9th January 2018.

8. Taylor M and Weatherspoon D. (2016) Deep Brain Stimulation (DBS). *Healthline.* https://www.healthline.com/health/depression/deep-brain-stimulation-dbs.

9. Phem U, Solbakk A-K, Skogseid IM and Malt UF. (2015) Personality changes after deep brain stimulation in Parkinson's Disease. *Parkinson's Dis* **2015**(13): 490507.

10. Griggs R. (2015) Coverage of the Phineas Gage story in introductory psychology textbooks: Was Gage no longer Gage? *Teach Psychol* **42**(3): 195–202.

11. *Alzheimer's Association* website: https://alz.org/dementia/types-of-dementia.asp

12. Maroon J, Winkelman R, Bost J, *et al.* (2015) Chronic traumatic encephalopathy in contact sports: A systematic review of all reported pathological cases. *PLoS One* **10**(2): e0117338.

13. Asplund C. (2015) Brain damage in American football. *BMJ* 2015, **350**: h1381 (editorial).

14. Asken B, Sullan MJ, DeKosky S, *et al.* (2017) Research gaps and controversies in chronic traumatic encephalopathy: A review. *JAMA Neurol* **74**(10): 1255–1262.

15. Di Virgilio T, Hunter A, Wilson L, *et al.* (2016) Evidence for acute electrophysiological and cognitive changes following routine soccer heading. *EBioMedicine,* **13**: 66–71.

16. Rutter M. (1981) Psychological sequelae of brain damage in children. *The Am J Psychiat* **138**(12): 1533–1544.

17. Swaine J. (2013) Brain damage fear hits junior American football. *The Telegraph.* 26th December 2013. https://www.telegraph.co.uk/news/worldnews/north-america/usa/10538152/Brain-damage-fear-hits-junior-American-football.html [18 October 2018].

18. *dementia.org* website https://www.dementia.org [18 October 2018].

19. Stryker S and Serpe R. (1982) Commitment, identity salience, and role behavior: Theory and research example. In: Ickes W and Knowles E (eds.) *Roles, Personality and Social Behavior,* pp. 199–218. Springer-Verlag, New York.

20. O'Connell T. (1998) *Making Disciples: A Handbook of Christian Moral Formation.* Crossroad Herder, New York.

21. Choudury S and Slaby J. (2016) *Critical Neuroscience: A Handbook of the Social and Cultural Contexts of Neuroscience.* John Wiley and Sons, New York.

22. Lieberman M. (2007) Social cognitive neuroscience: A review of core processes. *Ann Rev Psychol* **58**: 259–289.

23. Hjelmgaard K. (2017) Italian doctor says world's first human head transplant 'imminent' *USA Today.* Nov 20, 2017 (https://eu.usatoday.com/story/news/world/2017/11/17/italian-doctor-says-worlds-first-human-head-transplant-imminent/847288001/) [18 October 2018].

24. Leclercq S, Forsythe P and Bienenstock J. (2016) Posttraumatic stress disorder: Does the gut microbiome hold the key? *Can J Psychiat* **61**(4): 204.

25. Cryan J. (2016) Stress and the microbiota-gut-brain axis: An evolving concept in psychiatry. *Can J Psychiat* **61**(4): 201–203.

26. Carpenter S. (2012) That gut feeling. *American Psychological Association Monitor on Psychology,* September **43**(8): 50. (http://www.apa.org/monitor/2012/09/gut-feeling.aspx) [18 October 2018].

27. Salter M. (2016) Neurology and free will. *Iai.* (https://iai.tv/video/fate-freedom-and-neuroscience) [18 October 2018].

28. McCay A and Kennett J. (2016) Can neuroscience revolutionise the way we punish criminals? Retributivism is still alive and well. *The Independent.* 30 May 2016 (https://www.independent.co.uk/news/science/can-neuroscience-revolutionise-the-way-we-punish-criminals-a7056476.html) [18 October 2018].

29. O'Mara S. (2018) The captive brain: Torture and the neuroscience of humane interrogation. *QJM-Int J Med* **111**(2): 73–78.

30. Meynen G. (2013) A neurolaw perspective on psychiatric assessments of criminal responsibility: Decision-making, mental disorder, and the brain. *Int J Law Psychiat* **36**(2): 93–99.

31. Claydon L. (2012) Are there lessons to be learned from a more scientific approach to mental condition defences? *Int J Law Psychiat* **35**(2): 88–98.

32. Gkotsi G and Gasser J. (2016) Critique of the use of neuroscience in forensic psychiatric assessments: The issue of criminal responsibility. *L'Évolution Psychiatrique* **81**(2): e25–e36.

33. Spranger T. (ed.) (2012) *International Neurolaw: A Comparative Analysis.* Springer-Verlag, Berlin, Germany.

34. *Clark County Prosecution Office.* http://www.clarkprosecutor.org/html/death/US/slaughter955.htm [18 October 2018].

35. Schleim S. (2012) Brains in context in the neurolaw debate: The examples of free will and 'dangerous' brains. *Int J Law Psychiat* **35**(2): 104–111.

36. Markowitsch H and Staniloiu A. (2011) Neuroscience, neuroimaging and the law. *Cortex* **47**(10): 1248–1251.

8 Moving Beyond Mechanism in Medicine

William Beharrell*

A junior doctor in the NHS is required to rotate through a number of medical and surgical specialities. My last rotation was in haematology and, for some time after I completed it, my mind continued to return to the dozen or so patients that I attended in their last days of life; patients for whom treatments were ineffective, and whose lives were overcome by a haematological malignancy of one kind or another. As one of the junior doctors, I was expected to certify the deaths, and then, often, to examine the bodies in the mortuary prior to cremation. It was here, in this final encounter with a body, that I was confronted with the inadequacies of the models by which we medics interact with our patients.

In the cold hard surface of a dead body, there remain the many intricate impressions of a person, formed in relation to the world, from experiences accumulated over a lifetime. The integrating force, that served to unify this complex assemblage of parts into a single organism, and which somehow sustained a conscious awareness of this person's place in the world, dissipates in a process of autolysis and putrefaction. In dying, the person leaves a cadaveric ecosystem which is very much alive, host to entire microbial

*West Suffolk NHS Foundation Trust, Hardwick Lane, Bury St Edmunds 1P33 2QZ, UK.

communities, and unfolding into new forms of life. This is an essential process in the ongoing survival of humanity but it leaves us with questions about the nature of human life. How did this collection of matter generate life, and where has that life gone? What is it that makes this life distinctly human?

Rowan Williams, in a recent book review, suggested that one of the more persuasive answers to the question of how you could tell whether you were talking to some form of artificial intelligence rather than an actual human being was to ask them how they felt about dying. The medical team uses all the technical and pharmacological means at its disposal to cure the patient, but questions of meaning, fear, justice, forgiveness, and love remain, which if addressed in the right way, help reconcile the patient, and all those involved in their care to the reality of death.

One finds then, two sorts of language at the bedside: the language of biomedicine and the language of humanity. Clearly, they are not bound to be in competition, and there have been attempts such as Osler's 'medical humanism' to unite them. But it remains the case that these languages represent different perspectives on the body. It is important to recognise that they really are different discourses, making different assumptions, and can't just be fused together (any more than you can fuse language about the experience of sight with the processes that can be studied in the optic tract and cortex). Yet, neither perspective is fully adequate. If we are serious about the idea of 'patient-centred' medicine, it is crucial to work out how the human perspective relates to the broader western scientific account.

The central issue is how to negotiate a connection between a first-person subjective account of illness and a third-person objective account. Need this encounter be a negotiation? To what extent could we present a medical model that might equitably reflect the experiences and knowledge of both patient and doctor? Could such a model actually empower patients to take more responsibility for their own care? Does an approach, such as salutogenesis,[1] which focuses on factors that support human health and well-being rather than on what causes disease, deserve more attention in light of the current pressures facing the NHS?

There is scientific evidence to support the view that taking a human perspective enhances the effectiveness of medicine. For example, taking time and trouble to explain to people what they will experience in surgery reduces post-operative pain, speeds recovery and leads to earlier discharge. Though computers are better than doctors at combining complex diagnostic information, predictions are improved if the doctor's intuition is fed in as an extra piece of data.[2] This chapter is a call back to a more balanced way of thinking about bodies and being, and an approach to medicine that takes both fully into account.

Mechanism and the Machine Model

'We cannot speak of a machine "theory" of the organism, but at most of a machine fiction'[3] (p. 38)

Since early modernity, there has developed a view that the body is something akin to a machine. This metaphor has helped us to diagnose and repair dysfunctional parts, and has helped us to understand how many of the parts fit together. Yet this kind of reductionism does not account for a large part of human experience. And, it is easy to forget that it is, after all, only a metaphor. If we are not machines, what are we? An alternative view is that we are self-organising systems (though that might be taken as a broadening of the machine model rather than as a completely different approach). A self-organising system is associated with the idea of downward or whole-part causation, where the activity of the parts is influenced by the higher-level activity of the whole. This takes us away from a simplistic kind of monocausal determinism, and allows for multiple interacting causal factors. There is a strong scientific case for this, in medicine as in biology generally; many examples of the latter have been presented in this volume.

The value of a holistic approach would be widely recognised in medical science, but there persists a tendency, in practice, to focus exclusively on bottom-up causal factors to the neglect of the bigger picture. For example, when dealing with cancer, it is possible to focus exclusively on cancerous cells, and to neglect the regulative role of factors at the level of psychoneuroimmunology (PNI), which can be very important in healing and remission.[4] In recent decades, research in the field of PNI has elaborated multiple bidirectional pathways linking the central nervous system with the immune system. Studies suggest that downstream activation of the sympathetic nervous system initiates molecular signalling pathways involved in processes such as inflammation, metastasis and cell survival[5] as well as upstream bio-behaviourally modulated pathways that activate the parasympathetic system and affect tumour growth. It seems clear that there are multiple signalling pathways by which the 'macroenvironment' influences the tumour's 'microenvironment'. Psychological factors have been implicated both in the pathogenesis of disease but also in ameliorating the long-term unintended consequences of aggressive curative intent.[6] Given how important the immune system is to the process of homeostasis, it seems clear that any aetiopathogenic or prognostic account of illness should take into consideration the complex web of interactions between brain and body, but also the patient and their environment.

The notion that body and mind are separate, and perhaps even interchangeable was contained in what Descartes called the rational soul (*res cogitans*) and the body (*res extensa*), and of course expressed by the Cartesian idea that the body is a machine operated upon by the rational soul acting as efficient cause. This powerful idea informed the anthropologies of early modernity and ensured that the separation of our inward and outward lives remains dominant in the philosophy of mind to the present day. It also finds echoes in contemporary science fiction, artificial intelligence and post-humanism. In opposition to this view, the philosopher Mary Midgley refers to the 17th century essay on personal identity by Locke. Locke proposes that the soul of a prince could enter and inform the body of a cobbler and that everyone would see that he was the same person as the prince, accountable only for the prince's actions. She argues that the cobbler's mind needs the cobbler's body: 'It is not likely that two people with different nerves and sense organs would perceive the world in the same way, let alone have the same feelings about them, or that their memories could be transferred wholesale to a different brain...It's not at all like fitting a new cartridge in a printer, it's more like trying to fit the inside of one teapot to the outside of another'[7] (p. 175).

Here we have the idea of the body and mind as two substances, and an introduction to the machine model which has influenced the way we have come to see the body in modernity. Of course, the choice of model is important in any attempt at understanding, for it determines what we see. Constraints are necessary when faced with the sheer complexity of the world, and they allow freedom to interpret and to test propositions about the world. But whatever truth emerges from such a model it is important not to lose sight of the model itself. When considering the public understanding of medicine, it seems that the machine model continues to predominate, among those teaching and practising the profession, but also amongst patients. We go to hospital to get 'patched-up', 'fixed', to have parts 'repaired' or 'replaced'. Some of us go for an annual health 'MOT'. Patients regularly use phrases such as 'hard-wired' or 'genetically coded'. From our language alone, it seems we are convinced we are machines.

Such language is suggestive of a deeper attitude, which aims for total mastery of the body. One can imagine a range of reasons why this attitude might have taken root, not least the wish to alleviate suffering. This metaphor has come to the fore with the modern tendency to believe that things only have value in so far as they serve our needs. Classically, a human creature's will was secondary to the end of its desires. But since the 17th century, we have adopted a philosophy that sees cognition as potentially separable from the body, with a will that is free to identify and pursue its own ends. Furthermore, we have adopted a version of 'the scientific model' that regards the world as mechanistic and predictable; subject to natural laws which apply equally everywhere. This vision misconstrued the scientific process such that it does not leave much room for those qualities of imagination, wonder, intuition, and humility, which have characterised many of the great scientific discoveries. The machine model and 'the scientific method' have undoubtedly given us confidence to take charge of our own future, but as Bacon said: 'Nature to be commanded must be obeyed...The subtlety of Nature is greater many times over than the subtlety of the senses and understanding'[8] (pp. 47–48).

The notable biochemist Erwin Chargaff cautions us still further, 'In general, it is hoped that our road will lead to understanding; mostly it leads only to explanations. The difference between these two terms is also being forgotten...These are two very different things, for we understand very little about nature. Even the most exact of our exact sciences float above axiomatic abysses that cannot be explored. It is true, when one's reason runs a fever, one believes, as in a dream, that this understanding can be grasped; but when one wakes up and the fever is gone, all one is left with are litanies of shallowness'[9] (p. 56).

This puts us in mind, once again, of that central encounter between a doctor, who seeks to explain, and a patient, who seeks to understand. Writing with regard to the foundations of medical science, theoretical physicist Lee Smolin thinks that, 'those scientists who work on the foundations of any given field are fully aware that the building blocks are never as solid as their colleagues — and, a fortiori, the general public — tend to believe'[10] (p. viii). To describe science as securely rooted in fact (in contrast let's say to the public's view of politics), he says, is only a half-truth at best. Yet, there appears to be a conviction among many of the public akin to a 'religious belief in the conditional power of organised science, one that has replaced unconditional religious belief in organised religion...We have managed to transfer religious belief into gullibility for whatever can masquerade as science'[11] (p.109; p. 229).

It may help to take the public understanding of genes as an example. When learning about genetics at medical school, the first thing we were told is that no one really knows what a gene is, still less how to define one. But among the public, there is an understanding that we each have something like a genetic blueprint that maps the course of our lives. Children are taught that genes program biological processes in a manner analogous to computer hardware or indeed, that cells themselves express genetic programmes by turning genes on or off. The trouble is, according to Harvard geneticist Richard Lewontin, that none of this is true. 'DNA is amongst the most inert and non-reactive of organic molecules'.[12] Philosopher of biology Daniel Nicholson adds, that 'it is only in the presence of pre-existing cellular apparatus, that any talk of gene action can even make sense. And what is more, the origin of that cellular apparatus cannot be traced back to the genes'[13] (pp. 162–174).

The principle to note here is that genes cannot be understood in isolation from the wider organism. Lewontin argues for a conception of natural selection that is based on the idea of the environment and the organism actively co-determining one another. In this dynamic understanding of natural selection, it is interesting to reflect on how organisms manage to maintain stability whilst being in a process of change. To do so requires an understanding of homeostasis and takes us away from the machine model towards the idea of a self-organising system. So far, you might think, we have highlighted some of the limitations of the machine model and have begun to get to grips with the complexity of a living organism, but we are no closer to understanding where the human fits in. How do mind and body fit together and why, as Midgely asked, does the cobbler's mind need the cobbler's body?

Homeostasis and the Wisdom of the Body

'There is more sense in thy body than in thy best wisdom'[14] (p. 12)

The term homeostasis was introduced by the Harvard physiologist, Professor Walter B. Canon, who outlined his ideas for the relation of the autonomic nervous system to the regulation of steady states in the body in the Linacre Lecture in Cambridge in 1930.[15] He adopted the term, 'the wisdom of the body' from Professor Starling, who had himself expressed his admiration for 'the marvellous and beautiful adjustments in the organism'. 'Only by understanding the wisdom of the body' he said, 'shall we attain that mastery of disease and pain, which will enable us to relieve the burden of mankind'.[15]

The ability of living things to maintain their own constancy has long impressed biologists. Hippocrates (460–377 BC) implied the existence of agencies that are ready to operate correctively when the normal state of the organism is upset. The phrase that is generally attributed to him — *vis medicatrix naturae* — refers to the ability of the body to heal itself. The coordinated physiological processes which maintain a low-entropic steady state in the complex living organism rely on an exchange of energy between the organism and its environment. Here, at the core of one of the most fundamental characteristics of an organism, is the idea of interdependence and process.

But there remains a tendency within medicine to regard the body as static, autonomous, and separated from its environment. We do not tend to see the body in time, nor do we place much of an emphasis on the body's relationship to its environment. We rely on snap-shots of structures and look for clear chains of causation. The assumption seems to be that with the remarkable advance of imaging techniques, the body is transparent to the human eye. We look for clear biomarkers of disease in order to offer patients definitive proof of causation and to provide 'targets' for treatment. Nowhere is this more relevant than in haematology and oncology, where alternations to the signalling pathways in cells affect cell proliferation, motility, and survival. But these are not linear sequences, with a clear beginning and end. Rather, it appears that each sequence interacts with other dynamic sequences, at each step, and that far from being a chain of causation, it resembles a network where 'everything does everything to everything'.[16] This has been expressed in Dumont *et al.*'s so-called 'horror graph'.[16] Here, four cascades of five steps generate a possible 760 positive and negative interactions. This does not take into account the multiplicity of different isoforms of proteins at the different levels of the cascades, the multiplicity of effects of each intermediate in each cascade, the stimulation by a cascade of the secretion of extracellular signals, or feedback or feedforward controls within cascades.

Further instances of reciprocal action can be found by looking at the way an enzyme interacts with a receptor. Medical textbook diagrams describe a 'lock and key' mechanism, whereas it would now appear, that their interaction would be better described as a dance. Both enzyme and receptor actively accommodate each other. 'The receptor looks less like a machine and more like a ... probability cloud of an almost infinite number of possible states'[17] (p. 81).

It is astonishing to see how intelligently the body adapts to illness, and humbling to acknowledge how it compensates for often misguided iatrogenic interventions. To take an example from haematology, patients with chronic lymphoblastic leukaemia were offered, during the second half of the last century, a treatment called extra corporeal blood irradiation. Leukocytes are more sensitive to radiation than other cells. The theory was that if you took arterial blood from a patient's arm, irradiated it, and returned it to the arterial circulation, the white blood cell count (which is elevated in CLL) would drop. However, after three days, the white blood cell count returned to its previous level. The body had adapted to maintain homeostasis and the patient was living with a raised white blood cell count quite happily.

To take a more commonplace example, many of our cereals are fortified with ferrous sulphate. The theory goes that cereals are a readily available food source; iron-deficiency anaemia is the most common nutritional deficiency; therefore, food fortification is the most cost-effective way of addressing this global health problem. However, iron bioavailability from ferrous sulphate, combined with food processing, means that the body absorbs only a small fraction of this iron. Generally, if we suspect we might be running low on vitamins, hormones, chondroitin, dopamine or whatever it may be, we think that we can top ourselves up with added supplements. Equally, if we think we may have too much of something, like cholesterol, then we can reduce our intake to improve our health. However, these assumptions are shown time and time again to be either unhelpful, costly or even harmful.[18] We think that if we understand the mechanism, it will yield a technical fix. And often it does so in the most impressive ways, such as the developing technology driving machine perfusion in organ transplants. But we seldom think in terms of adjusting the other variables in life, be they relationships, lifestyle expectations, or self-identity, which may well give us alternative ways of dealing with illness.

In the context of cancer, what might be the implications of such an approach? It is generally understood, and even expected, among the public, that once a tumour is detected, if it cannot be removed, it is attacked with a barrage of cytotoxic drugs or radiation. Either the tumour is dissolved, or it is beaten into remission. Whilst the tumour may be weakened, hopefully destroyed, the wider organism is also weakened. Patients receive input from a dietician in recognition of the importance of strengthening the entire body, but I think we miss many opportunities to strengthen the body in such situations. For example, we put £432m towards cancer research last year[19] — a figure out of all proportion with the funding put towards other complementary services, such as psychological support or family counselling. Holistic treatments are the domain of complementary medicine, which unfortunately is often seen as a last resort when all other therapy has failed.

An Embodied Self-Organising System

Western scientific culture holds that we see our bodies as physical structures with lived experiences: two structures — outer and inner; two substances — body and soul. Francesco Varela's work on the *The Embodied Mind*, together with Evan Thompson and Eleanor Rosch offers one way of unifying body and mind.[20] His central theory of *autopoiesis* refers to a system capable of reproducing and maintaining itself: a self-organising system. The cognitive processes in this kind of system belong to the relational domain of the living body coupled to its environment. Cognition depends directly on the body as a functional whole and not just on the brain. The authors argue that physical self-organisation is the external aspect of cognition and that cognition is the internal aspect of self-organisation. Organisms set norms within which they define their goals and from which comes meaning.[20]

This autopoietic constitution of meaning produces consciousness, will, and thought and, in this regard, we are more meaning-makers than machines. Such autopoietic systems relate to their environments in a contingent way that is in accordance with the logic of their own systemic organisation. This leads to the associated theory of *enaction*, which attributes elementary cognition to these systems and understands behaviour according to a norm that is generated by the organism itself. In describing the movement of the bacterium *E. Coli* as an example, Thompson describes motion as having cognitive valance, helping an organism to operate meaningfully in its environment.[20] The idea that we enact a world on the basis of phenomenological experience and physiological requirements offers a clear alternative to the Cartesian dualism to which we referred in the introduction.

The idea that, instead of an independent external world accessed by a rational intellect, there is a dependent world, brought into being by embodied action, is reinforced by the experiential findings from the field of interoception. We can understand interoception as the physiological sense of the condition of the body — a homeostatic mechanism that combines the central autonomic network, with the emotional motor system and much of the limbic system also.[21] The concept of visceral sensory feedback to the brain has been around since the very early 20th century, but the discovery of a unique ascending pathway to the cortex in monkeys, that appears to integrate sensory activity from all bodily tissues, gave rise to a revised definition at the turn of the millennium and a growing interest in this field of study. Hence, a great deal of work has emerged in recent years implicating deficits in interoceptive awareness with a range of psychiatric illness, such as eating disorders, anxiety, and depression. For the purpose of the present discussion, the relevance of interoception is to be found in its integration of emotional and physical activity. It helps us to understand how ancient practices such as meditation have helped to regulate emotional experience and align body with mind. One example offered by Strigo and Craig is the cardiorespiratory synchronisation achieved through breathing exercises is associated with optimal tissue oxygenation and improved mood.[21]

The literature from the fields of embodied cognition and interoception helps to coordinate the actions of both body and mind in the context of a self-organising system. It offers an account of how meaning is produced relative to the norms established by that system and how that system maintains itself relative to its environment. But we know from our own experience that we are not purely reactive; we anticipate what the world is likely to be like and prepare ourselves accordingly. This ability is defined in the term allostasis, which refers to predictive self-regulation and the idea of 'stability through change'. This kind of adaptive behaviour, which requires a level of centralised control, is described in this context as downward causation and depends on hierarchical top-down processing. It includes the idea that consciousness is a higher-order emergent property, which has a causal effect on brain activity and our interaction with the world. It is an idea analogous to the animal-agent view of biologists which connects lower-level reactions and higher-level responsiveness. These concepts of downward causation and centralised control begin to converge on the idea of human agency but do not in themselves offer us a way of addressing the intentionality of human experience.

The Biopyschosocial Model

'I actually do cut out my mind when I construct the real world around me. And I am not aware of this cutting out. And then I am very astonished that the scientific picture of the real world around me is very deficient. It gives a lot of factual information, puts all our experiences in a magnificently consistent order, but is ghastly silent about all the sundry that is really near to our heart, that really matters to us...science sometimes pretends to answer questions in these domains, but the answers are very often so silly that we are not inclined to take them seriously[22] (p. 95)

'Biopsychosocial' (BPS) is a portmanteau term drawing together the various aspects of human experience into a whole. It emerged from the philosophy of pragmatism and general systems theory. The Pragmatists William James and C.S. Peirce developed the notion that we can only know the truth of things in practice and that all truth is therefore provisional. Whilst this placed an important emphasis on individualisation of treatment and made space for the patient's own values and beliefs, it led to an 'anything goes' kind of eclecticism.

General systems theory (GST), developed in 1925 by the biologist Paul Weiss, was incorporated by Walter B. Canon into his own work on physiology and homeostasis. But it was the German biologist Von Bertalanffy, writing in The American Handbook of Psychiatry in 1974, who did most to introduce the view that the whole is greater than the sum of its parts, to a medical readership. The psychiatrist Roy Grinker saw the potential in GST for explaining the multifactorial nature of mental illness and argued for holism is some areas while accepting reductionism in others. Unlike Grinker, Engel was not a psychiatrist, but a physician with an interest in gastroenterology. But it is Engel who formally laid out the biopsychosocial model in an article in Science in 1977 and thereby extended it beyond mental illness to the rest of the medicine. Engel emphasised treating the person not just the patient, the importance of psychosocial factors in illness, and the centrality of the physician — patient relationship.

Engel's critique of the biomedical model is summed up by Borrell-Carrio, Buchmann & Epstein[23] (p. 20) who remind us that biological derangements do not help doctors elucidate the meaning of the symptoms for the patient. Moreover, they argue that psychological variables are more important determinants of susceptibility, severity, and the periodicity of an illness that is generally acknowledged. We see evidence of this in those patients who adopt the so-called sick role but who do not necessarily show signs of biological derangement. Even if one considers the matter from the point of view of treatment rather than diagnosis, one must still credit the patient–clinician relationship as influential to the medical outcome, hence the success of the placebo effect in some situations. Finally, and in keeping with the 'Hawthorne effect' they remind us that patients are profoundly influenced by the way they are studied, and furthermore, scientists are influenced by their subjects.

One can see how the BPS model was a welcome counter-balance to a biomedical view of disease, which neglected the social, psychological, and behavioural dimensions of illness. For example, it encouraged a longitudinal rather cross-sectional view of illness in the context of the whole person and made space for patients' own beliefs and preferences. In elevating the principle of autonomy, a principle that continues to be enforced today, patients were free to insist on their own course of preferred treatment. Such autonomy sits uncomfortably alongside a commitment to evidence-based medicine, where a course of treatment may be offered in the patient's best interest, yet rejected on grounds of personal preference. Here, we return to the central encounter, which has become a negotiation between two different hermeneutic accounts of experience. The danger with the BPS model is that the social or psychological factors are given disproportionate weight at the cost of biological factors. For example, Ghaemi describes how the epidemiologist Davey Smith critiques the BPS by reminding us that *H. Pylori* was introduced as a cause of peptic ulcer disease in 1983. But the first hypothesis of bacterial association with peptic ulcer disease was put forward as early as 1875, with an antibiotic receiving a patent in 1961. Were the medical profession to have had a more singular focus on the biology, then something less than 100 years would not have been allowed to elapse before a breakthrough was made[23] (p. 59).

The BPS has an interesting history, well described by Ghaemi in *The Rise and Fall of the BPS*, but ultimately, it is less a model than an attitude, and as such, of limited value in explaining the complex interaction of multiple causes. It has not adequately been able to integrate the two perspectives of human experience and biomedicine and as a consequence, reservations remain among healthcare professionals who may regard it as impractical, irrelevant, and perhaps even harmful. My experience as a medical student was that many would recognise the validity of the following assessment from Ghaemi: 'The BPS model is harmless, innocuous, totally acceptable ... but probably totally ineffective. It is the obvious model to have emerged from a society that has grown used to dividing things into categories without ever putting them back together again; to espousing a view that sees no one choice as better than any other; and ultimately to prioritise materialism as the surest path to truth'.[23]

Dr. Iona Heath, a previous president of the Royal College of General Practitioners, highlights the limitations of any rational attempt to define the totality of human experience. Evidence is essential but always insufficient for the care of patients, she says. As doctors we should seek to relieve distress and suffering, and to this end enable sick people to benefit from biomedical science while protecting them from its harms. She reminds us that clinicians must see and hear patients in the fullness of their humanity in order to minimise fear, locate hope, and explain symptoms and diagnoses in language that makes sense to that particular patient, to witness courage and endurance, and to accompany suffering. No biomedical evidence helps with any of this, as Wittgenstein anticipated: 'We feel that even if all possible scientific questions be answered, the problems of life have still not been touched at all'[24] (6.52).

Returning once again to the central encounter between doctor and patient, Heath describes a rift running through every consultation and she points to the following tensions: disease versus illness, objectivity versus subjectivity, technical versus existential, population versus individual, utilitarian versus deontological, normative versus descriptive, numbers versus words, quantitative versus qualitative, reason versus emotion, science versus poetry. It is worth returning here to the central issue of how to negotiate a connection between a first-person subjective account of illness and a third-person objective account. Few would disagree that patients benefit from attempts by doctors to integrate a first-person and a third-person account of their illness — after all, this is the basis for the BPS model. But how far can these perspectives be integrated in medical science?

A distinction is generally made between medical practice, in which two potentially different agendas must be melded into a commonly agreed treatment plan, and medical science, which is, if we recall our quote from Schrödinger, characterised by the dispassionate pursuit of data. This distinction manifests itself within medical practice in the quiet conflict between advocates of evidence-based medicine and proponents of the BPS model. Yet, regardless of whether our work is with patients at the bedside or with petri dishes in the lab, we are tied by our senses. It is unclear how any scientist can claim that their first-person presuppositions have not influenced their work. Scientific theories, as Karl Popper describes them, are conjectural and not inductively inferred from experience; they are provisional in the sense that we can never finally prove them. Theories are contingent on a particular history, made up of particular personalities, policies, and politics. They are also contingent on the continual revision of what was formerly taken to be true.

Over and against this kind of experimental biology stands mathematical and theoretical biology, an approach that takes an abstraction of an organism and uses it to test mathematical models and theoretical analyses. Scientists are dwelling in abstract representations of the objects they are trying to study. This can be described by the term hypostatisation or the fallacy of misplaced concreteness, in which a hypothetical construct is mistaken for a real event or physical entity. One possible counter to this trend is provided by Wittgenstein's theory of aspect–perception in *Philosophical Investigations*,[25] which holds that one thing can be seen in multiple ways. The paradigmatic example is that of the duckrabbit. Just as we see both duck and rabbit, we construct subtle meanings for a single word depending on its context. Rather than having fixed meanings, words are dynamic tools, crafting subtly new meanings in each individual hearing according to previous experience and current context. This is an important insight to bear in mind given the tendency to see words as offering conclusive meaning to phenomena and hence the temptation to seeing the world only in one way. A third-person objective perspective, based on reason, and expressed with words, clearly plays a crucial role in scientific discovery. But it is vital to our understanding of true scientific endeavour, to acknowledge that many of the great scientists describe not reason but a kind of subjective insight as foundational to their work.

One example is Barbara McClintock, the foremost cytogeneticist of her generation and winner of the Nobel Prize in 1985. Her theory of transposition demonstrated the role of centromeres and telomeres and suggested that DNA could rearrange itself. She is described in her biography as having a special quality of seeing and is reported to have said that to see further and deeper into the mysteries of genetics requires 'a feeling for the organism'. Her willingness to attend to the variety and complexity of organisms and their relationships to each other led her to conclude that the genome is not a static entity but a complex structure in a state of dynamic equilibrium. It is subject to alteration and rearrangement in a process of learning from its environment. For McClintock, as for other scientists, reason and experimentation on their own are not sufficient. To quote Einstein, 'All great achievements of science must start from intuitive knowledge'[26] (p. 31). In McClintock's words, 'The ultimate descriptive task for both artists and scientists, is to ensoul what one sees, to attribute to it the life one shares with it, one learns by identification.'[27] (p. 306)

One way of addressing this rift in the consultation room is by acknowledging that each hemisphere brings about a recognisably different human world. According to psychiatrist Tim Crow: 'Except in the light of lateralisation, nothing in human psychology or psychiatry makes any sense'.[28] In this context, it is worth exploring these tensions by looking at the work that has been done on bihemispheric difference, most recently by Iain McGilchrist in his book *The Master and His Emissary*.[29] McGilchrist claims that there are two ways of being in the world: one allows things to be present to us in all their embodied particularity, with all their changeability and impermanence, and their interconnectedness, as part of a whole that is forever in flux. The other is to step outside the flow of experience, and experience our experience in a special way: to re-present the world in a way that is less truthful, but apparently clearer, and therefore cast in a form which is more useful for manipulation of the world and one another. The right hemisphere pays attention to the other, whatever it is that exists apart from ourselves. It is given life by the relationship, the betweenness. The left hemisphere pays attention to the virtual world that it has created, which is self-consistent but self-contained, disconnected from the other, making it powerful, but only able to know itself. Our disposition to the world and one another determines what we come to have a relationship with, rather than the other way round. The kind of attention we pay actually alters the world.

Conclusion

In 2000, the *NHS Plan* boldly stated: 'Step by step over the next ten years the NHS must be redesigned to be patient centred and to offer a personalised service... by 2010 it will be commonplace'. If we are to learn how to put the patient first we need a new way of seeing the patient. Will this require a new model of medicine or perhaps a recovery of an old model? What kind of model can integrate soma, psyche, and spirit and help doctors decide which aspect of a patient's experience to pay attention to? We have seen that in many respects, medicine has become much like engineering. We have got used to the idea that bodies, like machines, can be taken apart and put back together again. They can be improved. But living organisms, by contrast, are beings that we don't fully understand. Once we have taken them apart, we cannot put them back together again in their living form. They are formed of matter (itself a mystery), animated by a kind of energy which we cannot define, and ultimately recast as matter taking another form. Living organisms are dependent on their environment, exist in harmony with it, and survive because one generation gives way to the next. Machines are self-contained systems that exist for a purpose and are discarded when an improved model arrives. If living organisms belong in nature and machines belong to technology, then it seems that resolving the relationship between nature and technology will be an important step ensuring that patients remains at the centre of our healthcare system. Put differently, the challenge remains to establish a model in which a reductionist focus can be integrated within a holistic conception of human life.

Biologist Kriti Sharma[30] offers us a way forward in this regard with her theory of interdependence. Acknowledging that 'the network may be the ascendant metaphor of our time', she describes a 'causally complex world in which products depend on processes, processes depend on products, wholes depend on parts, parts depend on wholes, and living beings depend on one another for our lives'. She critiques the easy view of interdependence as no more than independence — independent objects interacting strongly, weakly, reciprocally, sequentially and so on — and shows that a theory of interdependence can be refreshingly simple, useful, and clarifying against the assumptions of reductionists who would argue the complexity of something makes it inaccessible until broken down into parts. To make sense of those parts requires a willingness to see them as things that are mutually constituted. Sharma urges us to consider things in interaction rather than seeing them in isolation. This emphasis on contingent existence serves the present discussion well by drawing together the ideas of emergent properties, homeostasis and self-perception, and moving us much closer to a model of medicine based not just on individual pathology, but on the interpersonal and environmental determinants of health and illness.

Sharma gives us an enlivening view of phenomena and a way of moving beyond the conflicting accounts of holism and reductionism: 'Just as the metaphors available for describing the phenomenon called a cell are not exhausted by its comparisons to a kind of animal, thermostat, chemical reaction, or standing wave, similarly the metaphors available for describing living systems generally are not exhausted by the comparisons to passive materials or active agents'. By reconciling the language of biomedicine and humanity within a unified framework of interdependence, it is possible to imagine an integrative therapeutic approach to the doctor–patient encounter. 'We do not need to separate the material world from the mystical, nor do we need to eliminate the material or the mystical from existence. The material world and the mystical world could be exactly the same place in every respect'. Such an approach would acknowledge the causal complexity of illness, within which a patient's subjective perspective may be just one of many contributing factors. Instead of buying into the animate–inanimate divide on the one hand, or panpsychism on the other, it would acknowledge that organisms and environments co-construct one another. We are other-relating and self-relating living organisms resistant to definition on our own terms, but continuously and radically dependent on each other and on non-living matter for our survival. Such an approach would call for respect for a patient's perspective not just because it might help elicit a diagnosis, but because it may be either a major determinant of illness itself, or just as importantly, may be central to that patient's own treatment.

We return here to the concept of salutogenesis, which was raised in the introduction as a potential response to the question of how to encourage patients to take responsibility for their own health. A mechanistic approach to the body tends to lead to technical solutions to problems of which there are simple linear causes. These 'top-down' solutions may depend on cognitive reevaluation, such as Cognitive Behavioural Therapy, or on pharmacological therapies to alter a particular chemical pathway. In both examples, the patients are expected to present themselves to a doctor in sufficient time for a therapy to work, and to receive that therapy as instructed. It is not hard to see why treatments are so often ineffective. For a variety of reasons, patients do not present in the right way or at the right time; they may not 'comply' with medication or medical advice. Insights from the fields of embodied cognition interoception and interdependence may be instructive for doctors and patients alike. If patients could be encouraged to appreciate the reciprocal relationship between mental and physical phenomena, then perhaps there might be an opportunity to promote an alternative view of the body — not as something which you 'own' and which is there for your own utility, but as something which is 'given' by others and which exists for others. Likewise, for doctors, might there be occasions when they see the patient not as someone whose parts needs fixing, but as someone whose need for healing requires relational and adaptive solutions prior to or alongside technical solutions?

In a new era of personalised medicine, in which new fields of medical practice are emerging such as pharmacogenetics, it is tempting to think that we are moving closer to a 'patient-centred' model of medicine. The reductionism that has, to a large extent, enabled these kinds of technical advance, cannot give an account of the distinctiveness of human life, and risks displacing patients from their rightful place at the centre of the medical model. A theory of interdependence offers to draw us back to seeing life in terms of cooperation rather than competition and has potential for integrating the various aspects of human experience. Together with embodied cognition and interoception, these theories place an emphasis on connection, process, and mutual dependence. Without such an account, our healthcare system will remain confined to a material understanding of life at the cost of being unable to recognise and respond to the complex causality of illness. Such complexity requires the humility to question the assumption that we know everything basic there is to know about matter and bodies, the wisdom to accept that not all technological advances are necessary or appropriate, and the courage to recognise that freedom comes not despite but because of our dependence on each other.

References

1. Antonovsky A. (1987) The salutogenic perspective: Toward a new view of health and illness. *Advances* **4**(1): 47–55.
2. Watts FN. (1980) Clinical judgment and clinical training. *Brit J Med Psychol* **53**: 95–108.
3. Von Bertalanffy L. (1933) *Modern Theories of Development: An Introduction to Theoretical Biology.* Oxford University Press, Oxford, p. 38.
4. Boivin MJ and Webb B. (2011). Modelling the biomedical model of spirituality through breast cancer. In Watts F (ed.) *Spiritual Healing: Scientific and Religious Perspectives.* Cambridge University Press, Cambridge.

5. Cole SW and Sood AK. (2012) Molecular pathways: Beta-adrenergic signaling in cancer. *Clin Cancer Res* **18**: 1201–1206.

6. McDonald PG, O'Connell M and Lutgendorf SK. (2013) Psychoneuroimmunology and cancer: A decade of discovery, paradigm shifts, and methodological innovations. *Brain Behav Immun* **30**(0): S1–S9. doi:10.1016/j.bbi.2013.01.003.

7. Midgely M. (2004) 'Souls, Minds, and Planets' Quoted in 'The Resounding Soul' Lee & Kimbriel (eds.) (2015) Veritas, Eugene, OR, p. 175.

8. Bacon F. (1620) *Novum Organon, in The Works, vol. 4.* Spedding J, Ellis RL and Heath DD (trans. and ed.), Longman, London, 1858, 39–248, Bk I, aphorisms III & X, 47–48.

9. Chargaff E. (1978) *Heraclitean Fire: Sketches from a Life Before Nature.* Rockefeller University Press, New York, p. 56.

10. Smolin L. (2008) *The Trouble With Physics: The Rise of String Theory, the Fall of Science and What Comes Next.* Penguin, London.

11. Taleb NN. (2012) *Antifragile — Things That Gain from Disorder.* Random House, New York 229, 109.

12. Lewontin RC. (2000). In foreword to Oyama S (ed.) *The Ontogeny of Information.* Duke University Press, Durham NC, 2nd edn, pp. xii–xiii.

13. Nicholson DJ. (2014) The machine conception of the organism in development and evolution: A critical analysis. *Stud Hist Philos Biol Biomed Sci* **48**: 162–174.

14. Nietzsche F. (2011) *Also Sprach Zarathustra.* Nikol Verlagsgesellschaft, Hamburg, p. 12.

15. Canon WB. (1962) *The Wisdom of the Body.* Norton, New York.

16. Dumont JE, Pécasse F and Maenhaut C. (2001) Crosstalk and specificity in signalling: Are we crosstalking ourselves into general confusion? *Cell Signal* **13**: 457–463.

17. Mayer BJ, Blinov ML and Loew LM. (2009) Molecular machines or pleiomorphic ensembles: Signaling complexes revisited. *J Biol* **8**(9): 81.

18. Illich I. (1988) *Medical Nemesis: The Expropriation of Health.* Random House, USA.

19. https://www.cancerresearchuk.org/ [31 December 2018].

20. Varela FJ, Thompson E and Rosch E. (2016) *The Embodied Mind [Revised Edition].* MIT Press, Cambridge.

21. Strigo IA and Craig AD. (2016) Interoception, homeostatic emotions and sympathovagal balance. *Phil Trans R Soc B.* **371**: 20160010.

22. Schrodinger E. (1954) Nature and the Greeks. In: *Nature and the Greeks and Science and Humanism.* (1996) Cambridge University Press, Cambridge.

23. Ghaemi N. (2010). *The Rise and Fall of the Biopsychosocial Model.* Johns Hopkins University Press, p. 20.

24. Wittgenstein L. (1994). *Tractatus Logico-Philosophicus.* Routledge, London 6.52.

25. Wittgenstein L. (1963) *Philosophical Investigations* (Anscombe G trans.) See Part 2, Section XI. Macmillan, New York.

26. Einstein A. (1931) *Cosmic Religion: With Other Opinions and Aphorisms.* Covici, New York, p. 97.

27. Keller EF and Mandelbrot B. (1983) *A Feeling for the Organism: The Life and Work of Barbara McClintock.* Henry Holt, New York.

28. Crow TJ. (2006) March 27, 1827 and what happened later — the impact of psychiatry on evolutionary theory. *Prog Neuropsychoph* **30**: 785–796.

29. McGilchrist I. (2009) *The Master and His Emissary: The Divided Brain and the Makings of the Western World.* Yale University Press, New Haven.

30. Sharma K. (2015) *Interdependence: Biology and Beyond.* Fordham University Press, Fordham.

9 The Reintegration of Biology, or 'Nothing in Evolution Makes Sense Except in the Light of Ecology'

Richard Gunton* and Francis Gilbert†

Introduction

'Nothing in biology makes sense except in the light of evolution', said Theodosius Dobzhansky in an address to the American Society of Zoologists.[1] Dobzhansky was arguing for the continued importance of organismal biology — including evolution — amid fascination with the increasingly prolific discoveries of molecular biology concerning DNA and its function within the cell. How does the perennial biological tradition studying 'how things are' sit with the Darwinian study of 'how things came to be this way'? Dobzhansky would no doubt be pleased to see the prominent place held by the notion of evolution in 21st century culture, and also reassured to know that many of today's general biology textbooks give prominence to the theory of evolution by means of natural selection. He would also be pleased to know about the new light being shed on genetic correlates and drivers of evolutionary change in molecular biology, including developments in the

*University of Winchester, Sparkford Road, Winchester, UK.
†Faculty of Medicine and Health Sciences, University of Nottingham, Nottingham NG7 2RD, UK.

study of epigenetics. But questions continue to be asked about what kind of scientific artefact evolutionary theory is, and to a large degree the challenge of relating evolutionary biology to molecular biology remains as grand now as it was then. Dobzhansky might be saddened at the continued decline in the teaching of systematics and taxonomy in leading universities and at what is probably a growing imbalance between coverage of molecular versus organismal biology at undergraduate level. Shedding light from evolution into studies of biochemistry, molecular genetics and cell physiology, and helping students appreciate the relationships between them, is perhaps as challenging as ever.

The idea we explore in this chapter is that the biological discipline of ecology is fundamental to the relationship between evolution and the rest of biology. Thus, we explore a three-way relationship between ecology, evolution and biology. We argue that evolution and ecology need to be taken as mutually interdependent, and that biology as a whole will benefit from a reintegration. Although ecology and evolution are parts of biology in its broad sense, we will use 'general biology' to mean the rest of the biological sciences apart from these two — and especially cellular and molecular biology. Ultimately, we ask what a more ecological version of biology — a reintegrated biology — might look like.

The Senses of Ecology and Evolution

Some initial light can be shed by looking at the range of meanings of the terms 'ecology' and 'evolution'. Our first term was coined by Ernst Haeckel, a friend of Charles Darwin, in 1866. From its Greek roots, 'ecology' literally means 'study of home', and it has always been used scientifically to describe the study of living organisms at home in their environments. This includes the interactions of individual plants and animals with other individuals that are closely related, as well as with members of other species; it extends from mutually beneficial relationships through to predation and parasitism. It also includes interactions of organisms with their non-living environments (like the atmosphere) and partially-living environments (like soil). Ecological science also considers patterns of organisms across space and how they change through time. In this way, as we move from timescales that can easily be studied in an ecological survey or monitoring programme towards timescales long enough for lineages to undergo changes in their characteristics, ecological science blends into its neighbour, evolutionary science.

We will look in more detail at the characteristics of scientific ecology below, but for now we will note two other meanings that have been derived from the scientific sense. First, 'ecology' is sometimes used to refer to a system of interactions among any kind of agents, as when economists talk of the ecology of a financial system. This seems to be a simple analogy, drawing upon the richness of ecological science in describing complex relationships among individuals — which need not be organisms. Second, in popular use 'ecology' has come to evoke political movements for sustainable living and environmentally responsible practices. In this ethical sense it is more often used as an adjective, 'ecological' being similar to 'green' in phrases such as 'the green movement'. This usage, which dates back to the 1970s, reflects how the study material of ecological science is under threat. Like archaeologists and anthropologists, ecologists study a diversity of things that can't be recovered if they are lost. 'Ecology' as a term, then, turns out be pregnant with wider meaning. It can provide a scientific analogy for other areas of thought, and it can point to a popular movement that transcends scientific study.

'Evolution' has just as diverse a range of meanings. Sometimes it is used simply to mean 'gradual development' or 'progressive change', as when we speak of the evolution of a language. In popular culture, 'evolution' can also stand for a materialistic view in which everything has gradually developed through natural processes out of previously-existing matter and energy in a grand cosmological history. Then there is the narrower scientific sense of evolution as a biological theory for the origin of species. Charles Darwin, in his culture-shaping book *The Origin of Species*,[2] suggested that the differences between biological species are not absolute but only a matter of degrees of continuous variation, disputing the classical view that species were real distinct types. And the more popular materialistic worldview resonates in some ways with Darwin's interest to show how the origin of biological diversity could be explained by laws of nature, just as ongoing planetary motion can be explained by laws of physics. This vision led to Darwin's most significant contribution to biological thought, and the central meaning of evolution. The process of evolution by natural selection, as envisaged by Darwin and his contemporary Alfred Russel Wallace, is one in which, as living things produce offspring that differ from their parents, certain changes progressively accumulate over many generations. This is evolution as a phenomenon, and gives its name to evolution the scientific theory.

In the rest of this chapter we will look in more detail at the scientific senses of 'ecology' and 'evolution' and their relationships with the rest of biology. What kinds of theories and principles are we dealing with? As we address this, we will uncover some intriguing and perhaps unexpected characteristics of the structure of science, as well as some fascinating idiosyncrasies about the biological sciences.

An Integrative View of Science: Law and Order

In order to account for the diversity of the different scientific activities that make up biological research, we need a broad definition of science. Science is commonly defined in terms of knowledge, in line with its etymological root, but this is at once too restrictive and too broad. It is over-restrictive because definitions of 'knowledge' are themselves subject to debate, as are questions of how 'true' or 'real' scientific concepts can be. At the same time, it is too broad because large swathes of 'knowledge' do not seem to be scientific: memories, acquaintance and tacit knowledge, for example. Instead, in line with our integrative approach, let's try the following: science is the human endeavour to describe the hidden rational order of the cosmos. 'Cosmos' is from the Greek word for order, which is appropriate enough, as we do indeed start with a conviction that the world is orderly and then seek to specify that order with increasing precision. From physics to sociology and from astronomy to linguistics, scientific work involves formal description of regular order: things like constant relationships, underlying structures and general systems that seem to recur across space and time. This is why scientific knowledge is often useful for prediction, explanation and devising new technologies: one place and time can be treated much like another once we discern some underlying order.

This simple description of science allows us to say what the diverse fields of ecology, evolution and biology have in common, and to outline their characteristic differences. First of all, ecology tends to be a search for order where there may appear to be none: in the apparent disorderliness of our natural environment and the haphazard encounters of organisms with each other. Ecological order sometimes yields to mathematical description only when ecologists try zooming out to observe large enough areas: it is characteristically a spatial science. Evolutionary science, on the other hand, is more like an attempt to order diverse categories of organisms by reconstructing their family trees on the basis of scientific reasoning — that is, by reasoning from other kinds of order. Evolutionary order refers to a past far beyond living memory but one that can be hypothetically reconstructed and ordered using other sciences such as genetics and geology. It is about zooming out, as it were, across different spans of past time so as to put the diversity of observed life forms into a meaningful order: typically what we call a phylogeny. Finally, we come to general biology. It is of course somewhat crude to lump the rest of biology together, but for present purposes we may portray biology as a search for layers of order and their interrelations as they contribute to the functioning of organisms (see the chapter by Leyser and Wiseman). This is at least sufficient to draw contrasts with ecology and evolution. In brief, we are suggesting that ecologists search for order by looking at different spatial scales, evolutionists find order in scientific hypotheses about the past, while other biologists, confronted with a great deal of order at the outset, describe how different levels of order relate to each other in the functioning of living organisms.

The biological sciences as a whole are a rich and diverse set of disciplines, practised by people with a wide range of interests and skills.

Order is a helpful term, but we need to dig a little deeper into the philosophical roots of the sciences to articulate an overarching perspective that will help us to reintegrate biology. If science is about describing the rational order of the cosmos as accurately as possible, it should be possible to say something about how biological 'order' can be described 'accurately', and what is meant by 'rational' here. We don't simply mean placing items in order by speculation or whim! One important and traditional notion that can take us further is that of laws: scientific laws, and laws of nature. Several important features of these concepts will help us think about reintegrating biology.

First, laws have figured prominently in the discourse of the natural sciences at least since Johannes Kepler,[3] who described a set of laws of planetary motion. Scientific laws have long been conceived of as laws of nature: inviolable prescriptions for how things must, and always do, happen, given certain conditions. Laws therefore allow us to make deductions: in the appropriate conditions, a certain phenomenon will occur. The conditions can be very stringent or even perhaps impossible, such as 'if there is no friction', 'if colliding bodies are perfectly elastic' or, in biology, 'if the environment is constant' (no weather! — Implausible but achieved by creating laboratory conditions), for example.

A second feature of scientific laws is that they are often stated in mathematical terms, as equations. This entails precision, and is one of the senses in which science may be said to describe a 'rational' order. This is especially interesting because the correspondence between mathematical structures and the material world has sometimes evoked surprise in scientists. Maths is developed from intuitions and axioms that do not refer to physical systems, so it seems remarkable when a piece of maths turns out to be applicable and useful for understanding a physical or biological system. In biology, we may think of the occurrence of numerical sequences such as the Fibonacci sequence when the numbers of repeating units in an organ are counted, like scales in a pinecone (other examples will be mentioned when we discuss laws in ecology in the next section). The 'unreasonable effectiveness of mathematics' for the sciences, as Eugene Wigner called it,[4] implies an order of correlation among different aspects of reality, and different sciences. The applicability of simple mathematical laws to scientific problems, and similarly of physical analyses to living organisms, and of biological approaches to psychological problems — such discoveries as these all contribute to the excitement of the scientific project of describing a rational order in the cosmos.

A third important point concerns the provisionality of scientific laws. For example, while Galileo's law of inertia has stood the test of time and is now better known as Newton's First Law of Motion, his law of planetary motion conflicted with Kepler's laws and has not been retained.[3] Scientific laws, clearly, are not beyond revision or rejection, and cannot be assumed to be real laws of nature. There is a large philosophical debate as to whether any law that a scientist might describe can refer to a real law of nature, as well as the question of how scientists could be sure of this. But it is clear that scientific views are subject to revision. This should come as no surprise in view of their speculative nature. Newton posited his laws of motion as universally applicable throughout all space and time despite his never having left the Earth and without being a time-traveller! These laws were duly celebrated and widely taught as laws of nature until Albert Einstein's theory of General Relativity relativised them as mere approximations to the results of laws described in terms of curved space-time. This is typical of the fate of scientific laws through history: they may be superseded, yet mostly survive major theoretical changes without being rejected outright. Now from our perspective early in the 21st century, Einstein's laws seem to be correct, having been validated to high precision by numerous experiments and observational tests — but the lesson we must learn from history, and perhaps intuition, is that scientific laws are always provisional. They seem to express laws of nature, but nature has a habit of proving more subtle than scientists expect.

We shall look at the scientific nature of ecology in more detail in the next section, but here we briefly consider the scientific characteristics of evolution. The principle of natural selection is that individuals whose traits make them more likely to survive to reproduce and then reproduce more successfully eventually leave more descendants, thus contributing more to successive generations, than those individuals with other characteristics. To have evolution by natural selection, it is necessary that traits subject to such selection are heritable — otherwise the mechanism is reset at each generation and nothing fundamentally changes — and that novel variation in selected traits continues to arise — otherwise change will cease or move in cycles. Understood in this way, evolution is the scientific study of regular processes occurring in the natural world in such a way as to produce certain kinds of patterns — or order — in the diversity of living organisms. It is important to note that we refer to evolution here as a process, and natural selection as a principle: neither are laws in the strict sense of empirical descriptions of either necessary or contingent order from which deductions can be made. Darwin might be thought to have discovered a law of biological evolution on the template of the laws of physics, as Haeckel suggested in applying Emmanuel Kant's phrase 'a Newton of the blade of grass' to Darwin. Indeed Darwin himself perhaps hinted at such an ambition in the closing sentence of the Origin of Species, which is worth quoting for its eloquent subtlety:

> There is grandeur in this view of life, with its several powers, having been breathed originally into a few forms or into one; and that, whilst this planet has gone cycling on according to the fixed law of gravity, from so simple a beginning endless forms most beautiful and most varied have been, and are being, evolved.[2]

But what law of nature did Darwin propose, if any? He refers earlier in the book to 'one general law, leading to the advancement of all organic beings, namely, multiply, vary, let the strongest live and the weakest die'.

This doesn't fulfil the standard notion of scientific laws outlined above: it does not allow predictions to be deduced and there is no conceivable way it could be falsified. Rather, it stands as a rule or algorithm that neatly summarises the principle of evolution by natural selection. In fact, we argue below that the scientific character of evolutionary science does not rest on any evolutionary laws but on the way in which it is situated in the context of other areas of biology — especially ecology.

This discussion of scientific laws is important because such laws appear in the different biological sciences in some contrasting ways — which together build up an integrated picture of biological order. We now need to look at the case of ecology.

Ecology and Its Laws

One of the striking things about today's science of ecology is just how successful ecologists have been at finding relationships between a diverse array of things that aren't easily measured or can't be detected by our senses at all. This is a facet of ecologists' search for order where naively we would see chaos. It is perhaps one of the reasons, incidentally, why ecology is under-appreciated in secondary education, since it is often necessary to use statistical techniques to interrogate large amounts of data before this order can be seen. Most ecological experiments cannot be done in a laboratory, nor are they particularly theatrical; rather they require data to be gathered from large geographical areas and/or long time-periods. But ecology is a science coming of age as ecologists appreciate the scope for unifying principles and theories. We are also becoming bolder in proposing laws.

We can begin with the numerical patterns that describe population dynamics. First comes the recognition that there are populations of individual animals and plants whose members interact with each other in a given region, such that changes in the size of these populations over time might show some kind of patterns. What size of region we should look at and how to sample it is one of the principal ecological challenges, but a fruitful answer to such challenges is revealed by the ability to describe numerical relationships between the size, or density, of a population from one season to the next, or between the size of a population of one species and that of another with which it interacts. Such descriptions have given rise to a class of laws known as density-dependence relationships. These laws generally take the form of simple mathematical models into which approximate numbers can be fitted to make sense of any particular situation, so they are not often useful for making precise predictions about the abundance of a species at a given time and place. Nevertheless, it was a crude precursor of such laws that inspired Darwin's theory of natural selection. Thomas Malthus' *An Essay on the Principle of Population*[5] pointed out that the numerical powers of increase of a human population were consistent with a geometric law which is clearly not fulfilled: if it were, the population size would increase at every generation by the average number of children born over the lifetime of each adult. Malthus suggested that the population growth would instead be limited by the food supply, which he whimsically suggested might be increased by a fixed amount at each generation. He contemplated a population growth rate that we would now describe as density-independent, and posited instead one that is density-dependent, constrained by the current density with respect to the food supply.

Another broad area of ecological laws is spatial laws. Spatial ecology has roots in the 19th century but took off following the suggestion that large groups of species might, in the wild, be functionally equivalent to each other. Such thinking seems to have begun with Robert MacArthur, who compared island faunas and asked how many species one should expect to find on different islands purely on account of the size of the island and its isolation from the nearest mainland. In an important way this was a continuation of Darwin's and Wallace's studies of the flora and fauna of archipelagos, yet the notion that different species might be treated as functionally equivalent was, in another sense, profoundly un-Darwinian. The unification of spatial ecology attained the status of a theory with Stephen Hubbell, whose unified neutral theory of biodiversity and biogeography[6] laid the foundations for a geometrical kind of ecology that has been remarkably successful in proposing ecological laws. The characteristic of these laws is that they describe numbers of species and other taxonomic categories to be found in areas of different sizes and distances from each other. They depend on evolution as a process of speciation, and in turn provide important underpinning for evolutionary work about diversification and adaptation.

Only a few attributes of species are treated by laws of spatial ecology, typically those relating to dispersal abilities and speciation propensities. The next area of ecological laws that we consider concerns species' attributes in all their complicated diversity. Functional trait ecology goes back to the roots of biological observations, as naturalists have speculated about the functional significance of species' traits. Since the late 20th century, however, advancing statistical capabilities and the accumulation of global data sets have facilitated an enormous growth in theorising about correlations among traits across different species. Why are most needle-leaved trees evergreen and lacking adaptations to spread their seeds far afield? Why do species of smaller birds tend to be shorter-lived and lay more eggs? Combining insights from evolutionary thinking with those from physiology — another of ecology's neighbouring disciplines — has allowed the emergence of relationships that should be considered as laws. The plant economics spectrum is a recent set of ideas that deserves to be popularised and may even constitute a theory. Laws have been discovered describing how adaptations that help plants conserve resources, like having needle leaves and being evergreen, tend to be associated with other adaptations that help them cope with more stressful environments, such as a tendency to grow in clumps of the same species.[7] The much older r-versus-K selection theory also provides a measure of law-like explanation — suggesting, for example, that species that produce offspring more prolifically tend to be shorter-lived — and so do the relatively new universal scaling laws,[8] which are part of a theory relating organisms' lifespans, body sizes and fluid transport networks. All these relationships are important for understanding both evolutionary and general biology: in fact they tend to provide a bridge between the two. General biology can often explain how traits come to be expressed in an organism and some of the physiological links between them; evolutionary biology explores how particular associations between traits are part of an organism's adaptation to a certain environment.

The final area of ecological laws we will look at is those pertaining to ecosystems. Given the prominence of the ecosystem concept in our culture, it is surprising to find that only a minority of scientific papers in ecology mention ecosystems. It also seems as though there has been less progress in discovering general laws that pertain to ecosystems, although plenty of measurements have been made of ecosystem properties. Quantities like biomass at different levels in a food web, energy fluxes between these levels, concentrations of different nutrients in the water, air and soil and in the bodies of organisms: all of these have been measured in different places and over timeframes ranging from minutes to decades — and indeed relationships have been detected. Some of the most general relationships may be the so-called resource-response models, with equations relating overall chlorophyll concentration, plankton biomass or primary productivity to the total phosphorus concentration of a lake, for example.[9] A further oddity about ecosystem ecology is how little it connects with the rest of biology — a point that we will reflect on later. Nevertheless, the ecosystem is the highest level of biotic phenomena we can recognise: a level that integrates all biotic functions and processes and connects them with non-living phenomena.

The four areas we have looked at are very different in terms of what kinds of phenomena they concern. From numbers of individuals to quantities of matter and energy, and from spatial patterns of organisms to their actual characteristics, very different kinds of prediction are possible, with applications in diverse areas from pest control to international policy, and from conservation planning to crop breeding. These four aspects of ecology still have much in common, however. All of them typically raise questions about spatial scale, and the patterns they seek are only detected with sufficiently large amounts of data gathered by zooming out to appropriate areas of observation — not too small, but not too broad either. All four aspects can also be related easily to considerations of natural selection and evolutionary change. We have hinted at some of the dependencies; we might now expect that, on closer investigation, different aspects of evolutionary theory are most relevant to these different aspects of ecology. It is, therefore, to evolutionary thinking that we now turn.

How Evolution Depends on Ecology

The scientific study of evolution is peculiar in several related ways. For one, it is difficult to make and test predictions. This is not a unique feature of evolutionary science: geology and sociology, for example, also weigh light on predictions, as do many other sciences. This is related to a crucial feature, that evolutionary science takes a special interest in the past. This historical focus is an important characteristic that will help us build an integrated view of the biological sciences.

There are other sciences that focus on the past, of course. Cosmology concerns, among other things, the development of the observed universe from a simple initial state — as in the Big Bang model. Geology explains the formation of the physical features of the Earth. Archaeology concerns the development of human civilisations. And evolutionary science concerns the history of life on Earth. Archaeological reconstructions, geological models and cosmological narratives are hypotheses, and one of the aims of evolutionary science is the construction of family trees. Perhaps the ultimate aim is a giant tree showing the relationships of all living organisms to each other — and to create this would of course require knowledge of deceased relatives. This cannot be done by direct observation or record-checking, as it often can with human family trees: it requires scientific analyses of markers of relationship, such as genetic sequences. And, by our definition of science given above, it must be the scientific nature of these analyses that give evolutionary trees their scientific interest. Merely positing an order among a given set of species is not a scientific achievement, but evolutionary scientists can do this on the basis of scientific reasoning.

Next, we should note that the processes studied within each of these historically-oriented disciplines have their origins in other, related scientific fields. Cosmology depends on the physics of fluids, energetic reactions and gravity; geology depends on various aspects of materials science; archaeology depends on diverse scientific theories in chemistry, biology, ecology, sociology and so on. So, what does the study of evolution depend on? Where do the various strands of evolutionary science find their home? We suggest that the scientific habitat for the web of evolutionary studies is the field of ecology.

Neo-Darwinian evolution is a very simple theory in essence, but an ecological context is needed to bring to life the beautiful tautology at its heart. The core principle of evolution by natural selection is essentially a logical one: among a set of things that can replicate themselves after their own kind, those kinds that replicate faster and survive longer will subsequently be proportionally more abundant. But to move from the resulting notion of 'survival of the fittest' to a theory of biological evolution calls for a biological model. First, we need a model describing inheritance with variation, that is, descent with modification. This much was provided in the neo-Darwinian synthesis of Dobzhansky with Ronald Fisher, J. B. S. Haldane and Sewall Wright, which drew upon the earlier work of Gregor Mendel. Inheritance was explained by the transmission of genetic material: DNA sequences being physically copied and passed into cells that give rise to offspring, while variation was mostly the result of random errors — mutations — in the copying process. Secondly, we need a model of natural selection. Using these two components it is easy to simulate evolution in a computer program — as has been done many times (e.g. 10). But such 'artificial life' simulations have rather contrived algorithms for selection. The choice over which entities to cull may be based on a range of criteria, but it is difficult to find many that relate to the world of living organisms. In some cases the filter even uses a human criterion of 'greater complexity' — introducing a teleological component that most biologists would strenuously resist. No: to understand evolution as a biological phenomenon, we need theories of ecology.

Ecology is a science whose roots are closely intertwined with those of evolutionary theory and the scientific community's gradual adoption of the ideas of Charles Darwin and Alfred Russel Wallace. Independently, and inspired by observations in different parts of the world, these two Victorian naturalists imagined how an enhanced survival rate for individuals with favoured characteristics could lead to gradual and potentially unlimited changes in the characteristics of a lineage. And this principle of evolution by natural selection helped turn scientific attention to the homes (habitats) of plants and animals, because if Nature were selecting favoured individuals and races, it must be doing so through the environments in which they eked out their living. It is, after all, with utter dependence on particular places and conditions that organisms live their lives, reproduce and die. So whether natural selection acts via direct competition, like two vultures fighting over a carcass, indirect competition, as when plants send roots into the same zone of soil, each extracting nutrients and water at the expense of others, or even apparent competition, as when an increase in numbers of one species bolsters the population of a predator and all the predator's other prey species then suffer as a consequence — in all cases the consideration of the environment is crucial for any understanding of the outcome. The environment is the *context* within which natural selection takes place, giving direction to the selective forces. Which animals happen to meet each other, which seeds land in proximity, which hapless individuals are washed or blown away — such random happenstances as these play a part in the overall course of evolutionary change. And ecology, as we have seen, is the science that seeks order in the juxtapositions and encounters, the accidents and serendipity, as they occur to living organisms. This science took root in the much older science-crafts of natural resource management: forestry, agriculture and fisheries,[11] and indeed Darwin himself contributed significant ecological studies of orchids[12] and of the earthworm.[13]

We may summarise by saying that evolutionary processes occur in an ecological context. But can ecological science complement evolutionary explanations?

Evolutionary Ecology

We noted in the previous section that merely putting a set of items in an order barely qualifies as a scientific activity unless it is done on the basis of some kind of scientific analysis. Evolutionary order, as in the family trees known as phylogenies, is scientific insofar as it is hypothesised on the basis of evidence of a scientific nature — using tools from fields such as genetics and geology — and explained in terms of regular patterns in birth, reproduction and death — which means ecology. Ecology is the crucial context for thinking about evolution because the characteristics of living organisms can only be interpreted in the context of their lives in some environment or other. Traits such as the beak of a bird and its length, or the fruit of a tree and its sugar content, are not absolute goods for organisms to have, or to have in greater measure: they can only benefit the organism in question by improving its survival and reproduction rates in the circumstances that it faces from moment to moment and from day to day. A trait that is advantageous in one year can easily be disadvantageous in another. To understand evolutionary patterns in traits, we therefore need to understand how those traits function in the struggle for existence — those interactions where individual organisms risk their lives or win reproductive success. Such interactions are the subject of ecology.

The importance of ecology for understanding evolution was for a long time obscured by fascination with the notion of inner forces of evolutionary progress. Indeed, the very term 'evolution', according to its etymology, means a rolling out, as if there might be a great predetermined chain of life-forms waiting to be revealed, like patterns on a rolled-up carpet. This pre-deterministic, law-like sense of 'evolution' is reflected in its use by chemists to mean the release of a gas in a chemical reaction: dropping sodium into a beaker of water causes the evolution of hydrogen, for example (and an explosion!). The neo-Darwinian paradigm, and evolutionary thinking in general, does of course take for granted an increase in complexity in the bodies and behaviours of many lineages of living organisms simply because of the assumption that the first living organism of all must have been very simple, and now we have myriads of species that are more complex. But this has never, to our knowledge, been successfully formulated as a testable law, or as the development of any kind of prescribed complexity — notwithstanding some fascinating debates about how far physical necessity may constrain the forms of complexity that we see.[14] The evolution of species is not a predictable kind of progress since it is difficult to find a sense in which it can be called 'progress' at all. Species that exist today are not 'better', 'higher' or 'more adapted' than those that have existed in the past. They are simply adapted to the context in which they find themselves today — a context that includes all the other living things they interact with in their communities. Communities of the past were very different from those of today, and hence the context of each species is also very different.

Small degrees of evolution are sometimes predictable, at least in laboratory conditions, where, for example, antibiotic resistance tends to arise in bacterial strains exposed to a particular antibiotic. Such resistance tends to be metabolically costly to produce, and hence tends to be lost when the stress is removed, and cannot be seen as part of a progressive evolutionary journey. More importantly, we must recognise that a phrase like 'selection will favour antibiotic resistance', while it may be a scientific prediction, cries out for an ecological context (the presence of an antibiotic) lest it should be taken as positing a teleological or animistic 'guiding hand'. Natural selection is a scientific metaphor for the outcome of diverse kinds of interactions between heritable traits and the environments in which organisms present them. It therefore needs an ecological explanation.

Ecologists love to find order in the contingent and the unpredictable: what happens when plants, animals and microbes are free in the wild to flee from or feed on each other, to escape fires or be struck by landslides, and so on. So, the importance of ecology in evolutionary studies has grown as the notion of evolutionary progress has receded. A tenet of scientific investigation is the continuity of underlying processes through time: scientists seek to discover processes and laws that are universally applicable at all times and places where relevant conditions and entities occur. This is where evolutionary ecology comes in. Evolutionary ecology is the study of how natural selection actually happens. It examines patterns in birth, reproduction and death rates: how it is that organisms do better or worse in the struggle for existence in terms of their heritable traits. Competition among individuals of the same animal species may occur in sexual (territorial behaviours, mating rituals, etc.) and non-sexual contexts (e.g. for food). Competition among plants of the same species occurs through competition below ground as root systems forage for nutrients and absorb water; it also occurs through competition for pollinating insects to produce the most abundant and fertile seeds. The results of such competitive interactions have been the subjects of ecological study for a long time, and now we have a pretty good idea of when they cause competitive exclusion, and when they have other effects, both direct and indirect.

We have said that ecologists study the haphazard and unpredictable encounters and fates of living organisms, which echoes the unpredictability of most evolutionary change. But earlier on we outlined a wide range of kinds of ecological laws that have been, and are being, discovered, and now we have suggested that this lawfulness of ecology actually makes an important contribution to the scientific status of evolution. This raises the thought that there might yet be more predictability in evolutionary change itself than is yet appreciated. With this in mind, we now turn at last to focus on the question of the reintegration of biology.

Making Sense of Biology: A Reintegration

In view of its prominence, we might suspect that evolution is seen as something more than a theory, even in scientific circles. Thomas Kuhn's celebrated book *The Structure of Scientific Revolutions*[15] provides an important concept that can help us make sense of the structure of biology and its major themes.

A paradigm, according to Kuhn and his interpreters, is a framework of accepted scientific problems and typical solutions, of textbook examples and approved methods, that characterises a particular field of research in a particular period, being recognised by the relevant scientific community as framing legitimate research in that field. It is, we might say, a framework of meaning for a programme of research. Kuhn's own examples of paradigms are drawn from the physical sciences — the paradigms of Newton's dynamic physics and of Einstein's General Relativity being two of the most comprehensive. But the application to biology is what concerns us here. The phrase of Dobzhansky with which we began this chapter suggests that he saw evolution as the reigning paradigm of the biological sciences: a framework in which everything in biology can make sense. In his address to the Zoological Society of America, Dobzhansky spoke of the twin themes of unity and diversity which the idea of evolution brings to the study of biology, and indeed the prominent place of evolutionary theory in biology textbooks and courses endorses the importance of this. But is there more to be said about how biology holds together? Let's notice that Dobzhansky's oft-repeated claim doesn't actually seem to depend upon the mechanism by which evolution occurs. The unity-in-diversity of living organisms is an insight arising from the principle that all living things ultimately arose from a single common ancestor (or it needn't be much weakened if we posited a few original life-forms instead of just one). But what about the principle of natural selection? Does that shed light on the whole of biology too?

At this point we could return to our earlier remark about 'evolution' referring to a materialist worldview, where indeed the principle of natural selection is heralded as a logical law with a creative force, which holds out the possibility of reducing biology to a kind of physics (see the chapter by Gatherer). But at this stage in the development of biology, and as evidenced by other chapters in this book, that reductionistic approach is not at all promising. Instead, we should look the other way. The principle of natural selection, we are suggesting, shows how evolution depends upon ecology. And there may be an ecological paradigm that sheds light on evolution itself.

To say that there are other paradigms besides evolution in the biological sciences is trivial. There are paradigms in developmental biology, in genetics, in plant and animal physiology and so on, each of which gathers communities of researchers, typical research goals and kinds of methodologies and data with which to achieve them, and specialised journals in which discoveries and ideas are published. And in ecology, the four areas of ecological laws mentioned above may be recognised as emerging from paradigms. We have a population paradigm, a spatial paradigm, a trait paradigm and an ecosystem paradigm. It is striking how much overlap and even cross-fertilisation among these paradigms there is: for example, there have been proposals of grand unifying theories of biodiversity that might sound like an endpoint for ecological research, but that turn out upon closer inspection to be laws concerning patterns in numbers of species as ranked from most to least abundant, or in numbers of species to be found in areas of different sizes and separation distances.[16]

But what kind of paradigm might give meaning to natural selection? We have noted the fallacy of taking natural selection as a 'guiding hand' — on the model of human agents engaged in the artificial selection of animals and plants to breed from in seeking to improve varieties for agriculture, for example. And we have mentioned that it is by means of ecological interactions that natural selection actually happens. Perhaps the most important way to understand natural selection is by considering the ultimate significance of the traits of organisms. Ultimately, our interest in evolution is in how species and other groupings come to differ from each other and to do different things — and this means understanding their traits. Traits, moreover, are in general the features of organisms that must be referred to in any account of ecological interactions concerning individuals. Functional trait ecology, then, is the paradigm that focuses on finding order among disparate individuals by measuring features that they have in common and seeking to understand the adaptive significance of these. There are questions of how different traits in a species relate to each other physiologically — what Darwin referred to as laws of growth, which he recognised would mould the effects of natural selection up to the point where such laws might change under selective pressure. There are the all-important questions of how particular traits affect the survival of individuals in conditions and circumstances that threaten them — from before birth through infant and juvenile stages to the period of reproductive activity, and including dispersal phases. There are also important questions concerning traits involved in sexual selection and other behavioural traits or tendencies that mediate survival and reproductive success. All of these are the subject of functional trait ecology.

To Dobzhansky's famous adage, therefore, we should add that nothing in evolution makes sense except in the light of ecology, and in particular the ecology of species' traits. Ecology, in turn, makes no sense apart from an understanding of the organismal and cellular fields of biology that account for the structures and functions of organisms which are manifested as traits. There is no foundational discipline among the biological sciences (Fig. 9.1).

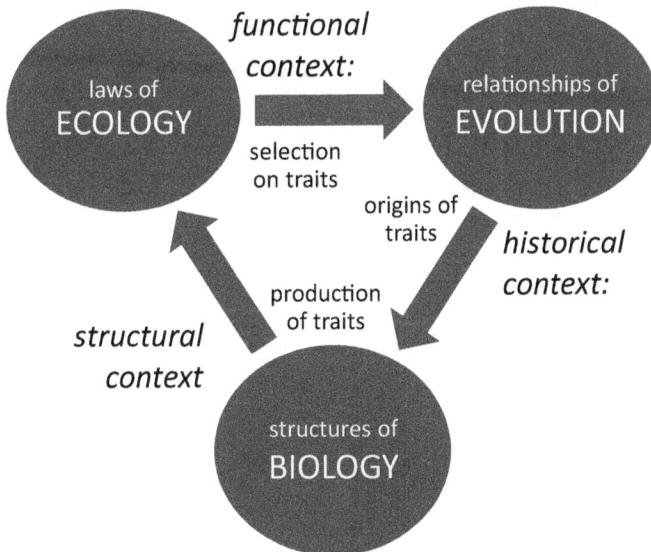

Figure 9.1. A model for the interdependence among the biological sciences. Ecology, evolution and general biology depend on each other, representing complementary facets of scientific thinking, and each providing an indispensable context for the others.

Outlook

How might people's understanding of biology be enhanced if ecology were properly understood as an indispensable link in the circle of life sciences? A reintegrated view of biology could bring benefits as far-reaching as improvements to our health, our family and social life and the impacts of our lifestyles on the environment. A less reductionistic and more ecological view of biology should, for example, affect our approach to diet, mental health, medicine, aging and long-term healthcare — some of which are explored in other chapters of this book. It could also directly affect our attitudes to nature conservation, land management and animal husbandry, as we appreciate more deeply the inter-relationships among living organisms.

We should also consider a reintegrated biology curriculum in education. A more ecological approach to evolution and a more evolutionary approach to ecology could enliven the teaching of both subjects. Evolution arguably needs to be better grounded in a geographical approach, which would encourage students to ask where natural selection may actually be happening — which calls for ecological insights into the various processes of competition. Ecology in turn needs to be better grounded in an understanding of how biology shapes and constrains the production of actual traits in plants, animals and other organisms. This should encourage students to ask how ecosystems might be different, and how emergent ecosystem processes like food webs and nutrient cycles reflect selective pressures on individual organisms — also helping locate ecosystem ecology with respect to other paradigms. Ultimately, we might see biology textbooks with a more integrated structure. They might be shorter rather than longer, and students might engage with biology less as a subject about complicated facts, and more as a discipline of ecological thinking.

References

1. Dobzhansky T. (1964) Biology, molecular and organismic. *Am Zool* **4**: 443–452.
2. Darwin CR. (1859) *The Origin of Species by Means of Natural Selection or The Preservation of Favoured Races in the Struggle for Life.* 1968 ed. Penguin Books, London.
3. Stafleu MD. (1987) *Theories at Work: on the Structure and Functioning of Theories in Science, in Particular During the Copernican Revolution.* Toronto: University Press of America.
4. Wigner EP. (1960) The unreasonable effectiveness of mathematics in the natural sciences. Richard Courant lecture in mathematical sciences delivered at New York University, May 11, 1959. *Commun Pur Appl Math* **13**(1): 1–14.
5. Malthus T. (1798) *An Essay on the Principle of Population as it Affects the Future Improvement of Society.* J. Johnson, London.
6. Hubbell SP. (2001) *The Unified Neutral Theory of Biodiversity and Biogeography.* Princeton University Press, Princeton.
7. Reich PB. (2014) The world-wide 'fast-slow' plant economics spectrum: A traits manifesto. *J Ecol* **102**(2): 275–301.
8. West GB, Brown JH and Enquist BJ. (1997) A general model for the origin of allometric scaling laws in biology. *Science* **276**(5309): 122–126.
9. Peters RH. (1991) *A critique for ecology.* Cambridge University Press, Cambridge.
10. Dawkins R. (1986). *The Blind Watchmaker.* Norton, London.
11. Bowler PJ. (1992). *The Fontana History of the Environmental Sciences.* Fontana, New York.
12. Darwin CR. (1862) *On the Various Contrivances by Which British and Foreign Orchids are Fertilised by Insects, and on the Good Effects of Intercrossing.* John Murray, London.
13. Darwin CR. (1882) *The Formation of Vegetable Mould Through the Action of Worms: With Observations on Their Habits.* John Murray, London.
14. Conway Morris S. (2003) *Life's Solution: Inevitable Humans in a Lonely Universe.* Cambridge University Press, Cambridge.
15. Kuhn TS. (1970) *The Structure of Scientific Revolutions,* 2nd edn. University of Chicago Press, Chicago.
16. McGill BJ. (2010) Towards a unification of unified theories of biodiversity. *Ecol Lett* **13**(5): 627–642.

10 Food and Nutrition

Michael J. Reiss*

Introduction

What we eat is core to our health, how we understand ourselves and a substantial part of our enjoyment in life. Yet today's diets differ greatly from those that we evolved to have. Furthermore, public understanding of nutrition is not strong and, for many people, there are major mismatches between what they would like to eat and what they actually consume, both in terms of amount and quality. In this chapter, I begin by describing the diets that were typical in our evolutionary past, explaining how they differ from those that are increasingly common today. Then, I examine what is meant by 'a good diet', broadening the notion from a narrow focus on physical health. I also look at how official advice about diet is sometimes less secure than is admitted and examine the current state of knowledge about fats and carbohydrates in our diet. I conclude by examining what good education about nutrition might entail both in schools and for adults, considering the role of food in identity and culture as well as in nutrition.

---

*Professor of Science Education, UCL Institute of Education, University College London, 20 Bedford Way, London WC1H 0AL, UK.

The Diets We Evolved to Eat

Humans, in common with the other great apes,[1] are omnivores. By this is meant that we have an impressive ability to consume a very wide range of organic matter and build ourselves out of it. Until we developed agriculture some 10,000 years ago, we got our food by gathering plant matter and hunting animals — mainly mammals, birds and fish, though some communities also ate other foods, such as molluscs (e.g. oysters). First people — and there are precious few of them left nowadays — who rely on food that is available in the wild and eat few or no domesticated plants or animals are typically able to identify hundreds of species in their environments, in large measure because they need to know which potential food species are edible and which are not.

Over the past few decades, there have been a spate of books advocating 'palaeo-diets'. Perhaps the best known is Loren Cordain's *The Paleo Diet*,[2] which on its front cover proudly states 'Lose Weight and Get Healthy by Eating the Foods You Were Designed to Eat', along with 'Over 100,000 copies sold!'. The popularity of such approaches stems from a growing dissatisfaction for many of us with how we feed ourselves. However, there are a number of problems with the palaeo-diet approach, and it certainly is no panacea. For a start, the average person in the Palaeolithic (from about three million to 15,000 years ago) was probably less overweight than many of us are, through a combination of more exercise and/or less energy intake. Then there is the issue that it is quite difficult — and certainly expensive — to get a diet that is close to what our ancestors were eating at this time. In particular, domesticated animals (bred since the end of the Palaeolithic) have a lot more fat than wild ones do, while domesticated plants (also bred since the end of the Palaeolithic) are a lot more digestible than are most wild plants. In addition, certain foods that are a standard component of most people's diets nowadays — e.g. cereals and legumes — barely featured then. The fruits were smaller and tarter, and potatoes (restricted to South America) were the size of today's peanuts. A palaeo-diet eschews alcohol, tea, coffee and chocolate too. Furthermore, the ancestors of some of our domestic plants can be quite dangerous — for instance, wild almonds produce dangerously high levels of cyanide.

Having said all that, most of us have diets nowadays that are higher in salt and sugars and lower in fibre than did our ancestors, so that palaeo-diets do have certain features to commend them. The fact that today's environment is so different from that in which we spent much of our evolutionary past causes problems with regards to the wisdom of our food preferences. As is well known, most people have a greater liking for saturated fats and for sugars than is healthy. This can cause problems for those of us who now live with a superabundance of foods rich in such ingredients. In the past, of course, there was little to be gained — and much to be risked — in not stuffing ourselves when an occasional glut of such foods presented themselves.[3]

Another issue is that human evolution hasn't stood still over the last 15,000 years, about 600 generations — enough time for quite a lot to happen in evolution. The classic example is the ability to digest lactose. The large majority of us who come from populations where milk from domestic cows (or other mammalian herbivores) is regularly consumed are able to digest this natural sugar whereas there is no doubt that, ancestrally, extremely few humans could.

Virtually no human societies have ever been exclusively vegetarian or carnivorous — though, of course, many people live their lives as vegetarians and the Inuit have a diet that is, on average, some 99% carnivorous. The one distinctive thing about the human diet, relative to that of other species, is that, for a mammal of our size (typically around 50–100 kg for an adult), it is of high quality. Most land mammals our size eat a much higher proportion of mediocre quality vegetation than we do.[4] The reason we need a higher quality diet is because of our brains. At rest, about 20% of our energy expenditure is on our brains, even though their mass is only about 2% of the body's. This high energy need means that we cannot subsist, unlike, for example, sheep, cattle, horses, gorillas or chimpanzees, on large amounts of fairly low-quality vegetation. Of course, many people have a diet that is largely or entirely plant-based — but that is predominantly based on seeds, fruits and nuts, not leaves and stalks.

What is a Good Diet?

The phrase 'a good diet' may seem clear but we need to ask two questions: what is meant by 'good' and good for whom? As I will attempt to show, these are not just academic, nit-picking considerations but fundamental to the question of public education about food and nutrition.

The simplest answer to the question 'what is meant by good?', when we speak of a good diet, is that such a diet is good for our health. Implicit within this is probably 'physical health' and there is much to be said about good physical health (consider the alternative). But to focus only on physical health in the context of food is a rather narrow way to envisage eating. For a start, food often plays a key role in companionship. If I eat rather more than might strictly be good for me (in terms of physical health) or even consume more alcohol (within reason) than is medically ideal when in the company of friends or family at a celebratory event, such as a birthday or wedding, am I unwise to have done so?

So, one way of understanding the phrase 'good diet' is to understand it as meaning a diet that helps me feel positive about myself, one that promotes my wellbeing (understood as including but being more than physical health) and contributes to flourishing, both for me and, when eating communally, for others. By and large, such an understanding corresponds to that of physical health — but it extends it to encompass broader aspects of health, including cultural aspects.

As far as determining what a diet that is good for physical health should consist of, there is still much disagreement among experts. As the astronomer Martin Rees puts it:

> Today, astronomers can convincingly attribute tiny vibrations in a gravitational wave detector to a 'crash' between two black holes more than a billion light years from Earth. In contrast, our

grasp of some familiar matters than interest all of us — diet and child care, for instance — is still so meagre that 'expert' advice changes from year to year. When I was young, milk and eggs were thought to be good; a decade later they were deemed dangerous because of their high cholesterol content; and now they seem again to be deemed harmless.[5]

Before birth, of course, one's diet is provided though the bloodstream of one's mother. Problems can arise if a pregnant woman's diet is high in alcohol (which can lead to fetal alcohol syndrome) or low in folate, vitamin B9 (which can lead to neural tube defects). Both of these can result in very serious, lifelong conditions. While fetal alcohol syndrome has a number of clinical manifestations, the most important is brain damage to the developing baby, with consequent neurological impairment. The most severe neural tube defects include spina bifida (when the spinal cord and spinal column don't completely close) and anencephaly (greatly reduced brain size). The problem of a pregnant woman not getting enough folate in her diet is compounded by the fact that this is needed in the first two to three months of pregnancy, before a woman may even be sure that she is pregnant.

After birth, by and large, human milk is best for the first year or two. It is hardly surprising that human milk is an ideal diet for a developing baby — it's what we have evolved over tens of millions of years to survive on for the first couple of years after birth. In modern societies, bottle milk can be just about as good from a nutritional perspective (though even that is doubted by some), but breastfeeding has additional benefits. It helps protect against certain common childhood infections and even lowers the incidence of Sudden Infant Death Syndrome (cot death). It may help reduce the risk of asthma and certain allergies, though this is much less certain. One can't really do controlled experiments in this field so all the studies tend to be correlational ones which introduces the problem of confounding variables (as it's hardly a random choice for a woman as to whether or for how long she breastfeeds her infant). Finally, many mothers enjoy breastfeeding — though it can be very painful for some — and this enjoyment, along with the physical contact with the mother, can communicate itself to the baby, with subsequent benefits.

Once the infant or young child moves away from depending exclusively on milk, the golden rule seems to be a balanced diet and moderation in all things, with neither too little nor too much overall. A balanced diet doesn't mean that we need particular foods; rather, it simply means that a range of foods is likely to provide us with the nutrients we need. We have evolved to be omnivores and one of the advantages of that is that if someone doesn't eat meat or eggs or dairy products or peanuts or whatever, or if certain foods become unavailable, chances are they will be fine so long as they continue to get a range of other foods. Admittedly, avoiding eating all plant matter is problematic; as a boy, my refusal to eat vegetables meant that I managed to develop scurvy, which was pretty embarrassing for my parents as they were very well off, with one of them a surgeon and the other a nurse.

Aside from the risk of developing scurvy (through vitamin C deficiency), a 100% meat diet (while possible — consider the Inuit, though the Inuit do eat a certain amount of plant food, as well as seaweed, and it needs to be remembered that their meat diet includes a lot of fish) has a number of problems, despite not infrequent over-enthusiastic celebrity endorsements. The nutritional problems with a diet that is close to 100% meat include shortages of fibre (part of the reason why a human carnivore is more likely to develop colorectal and other cancers), vitamin E and folate. To a lesser extent, at the other end of the spectrum, a vegan diet, for all that it can have a number of health advantages, can also be difficult to manage. Care may be needed to obtain enough protein, including all the essential amino acids, as well as vitamin B_{12} (generally taken as a supplement by vegans in wealthy countries).

More detailed advice as to what constitutes a diet that is good for physical health remains surprisingly contentious, as instanced by the Martin Rees quotation cited above. For example, most of us are aware that dietary advice about fats and carbohydrates has changed in recent years. A recent major review of the nutritional benefits and harms of different sorts of fats and carbohydrates, undertaken by a most reputable group of academics and published in *Science*, begins by noticing that since the mid-1970s, people in the US (and a number of other countries in the West) have had over thirty years of governmental advice that has consistently called on us to reduce our fat consumption, particularly saturated fats, and to increase correspondingly our carbohydrate consumption.[6] As a result, the percentage of fat in the typical US diet decreased from 42% in the 1970s to 34% in the 2010s. At the same time, the anticipated health benefits have not occurred. Indeed, obesity and diabetes rates have increased markedly.

Of course, lots of other things have changed in the US over this period. If we restrict ourselves to changes in diet, portion sizes have grown, as has eating out and the use of food processing. Non-diet changes include marked decreases in physical activity. The authors of the *Science* review summarise current diet controversies over fats and carbohydrates by posing a number of questions to which we do not yet have firm answers. Their questions include the following:

- Do high-fat, low-carbohydrate (ketogenic) diets provide health benefits beyond those of moderate carbohydrate restriction?
- What are the optimal amounts of specific fatty acids (saturated, monounsaturated, polyunsaturated) in the context of a very-low-carbohydrate diet?
- What is the relative importance for cardiovascular and other diseases of the amounts of LDL (low-density lipoprotein) cholesterol (so-called 'bad' cholesterol) and HDL (high-density lipoprotein) cholesterol (so-called 'good' cholesterol) in the blood?

At the same time, there are some issues where a consensus exists, always accepting that just because a consensus exists doesn't mean that knowledge is certain:

- Good health can be achieved for many people on diets with a broad range of carbohydrate-to-fat ratios.
- Replacement of saturated fat with naturally occurring unsaturated fats provides health benefits for most people. Industrially produced trans fats are harmful.

- Replacement of highly processed carbohydrates (including refined grains, potato products, and free sugars) with unprocessed carbohydrates (non-starchy vegetables, whole fruits, legumes, and whole or minimally processed grains) has health benefits.
- People with relatively normal insulin sensitivity and β cell function can do well on diets with a wide range of carbohydrate-to-fat ratios. Those with insulin resistance, hypersecretion of insulin or glucose intolerance may benefit from a lower-carbohydrate, higher-fat diet.
- Well-formulated low-carbohydrate, high-fat diets do not require high intakes of protein or animal products. Reduced carbohydrate consumption can be achieved by replacing grains, starchy vegetables and sugars with non-hydrogenated plant oils, nuts, seeds, avocado and other high-fat plant foods.

Interestingly, there is now evidence that having a diet that is either very low or very high in carbohydrates is associated with greater mortality than one that is intermediate.[7] Once again, moderation is not a bad guide.

The second question I said we need to ask is 'good for whom?'. So far, my emphasis has been on humans, mainly at the individual level of the person doing the eating. At a larger scale, food raises other issues to do with 'the good'. To take a cheerless example, the considerable rises in obesity rates have corresponding costs to national health services. It would be better, simply from the perspective of national economics, if fewer of us were very overweight. From the standpoint of political philosophy, this provides a justification for governments to attempt to influence what we eat. There is an analogy here with compulsory seat belts. In a country with a national health system, given that seat belts not only save lives but reduce government health expenditure, a government can claim the right to make seat belts mandatory.

But, of course, it's not just humans for whom our diet can be good or bad. Let's start with a relatively non-contentious though still distressing issue, the consumption of meat from domesticated animals when such animals have a life that is barely worth living. I am thinking of the (literally) tens of billions of chickens, cows, pigs and rabbits that are raised and slaughtered for food across the world each year. To give just one example, broilers are chickens raised for meat production. They have been bred to grow so fast that they typically live only about 35–50 days before they are slaughtered, compared to several years for their wild ancestors. Unsurprisingly, this huge increase in growth rates has been accompanied by skeletal and locomotory problems; the majority of broilers find walking painful. They live their lives under artificial light regimes (some in complete darkness, some in complete light, some in perpetual, Hadean gloom) at abnormally high densities and typically in rearing sheds that hold thousands or tens of thousands of birds. Catching the birds (for transport to slaughter) is traumatic and not infrequently leads to further injury.

The emphasis here is on animal welfare — and the lack of it that too many farmed animals experience. Other issues about goodness that our food raise relate to our use of land. For example, we continue to clear huge areas of natural vegetation, including tropical rainforest, with devastating consequences for global biodiversity, to plant soybean to feed to cattle to provide us with beef; these cattle produce large amounts of methane, contributing disproportionately to global climate change. In the seas, about one third of world fish stocks are over-fished.[8] I could go on.

What Might a Good Education about Food and Nutrition Entail?

It is quite difficult to assess public knowledge about food, given the considerable disagreements, as pointed out above, that exist among experts as to what constitutes good nutrition. Nevertheless, we can identify a number of components that together would go towards providing a good education about food and nutrition, one that would facilitate public understandings about these issues. I start with conceptual knowledge about nutrition — which is what school nutrition education normally consists of. But then I argue that this is not enough. Good education about food and nutrition also involves learning about the nature of science, about moral philosophy and about identity and culture. We need to rethink food and nutrition and we need to rethink education about food and nutrition.

Conceptual Knowledge about Nutrition

The amount of conceptual knowledge that school children or adults may need about nutrition is probably quite small. Byrne and Grace[9] suggest that in primary schools, children may only need to be taught about the importance of a balanced diet. Having said that, a balanced diet doesn't come cheap; in the UK, research by the Food Foundation[10] has shown that the costs of healthy eating fall disproportionately on those with limited resources. A healthy balanced diet for the poorer half of the population would account, on average, for close to 30% of disposable income; this compares with an average of 12% of disposable income for the wealthier half of households. Internationally, food inequalities are even more apparent.

In secondary schools, 11–14 year olds would benefit from being taught basic elements of digestion (including how tooth health can be maintained). By the time students reach the compulsory school leaving age (around 15 to 17 years in most countries), they should also know: the function of different food groups in maintaining health; the consequences of imbalances in the diet, including deficiency diseases and obesity; the calorific value of different food groups; and the requirements of a healthy, daily diet. Practical work in school science can be used to help reinforce much of this conceptual knowledge. For example, there are numerous experiments that show the action and specificity of digestive enzymes, while the use of simple calorimeters to show the calorific value of different foods is always engaging and food tests can be undertaken to help identify the presence of various food groups in foods. There are also software packages that enable students who have kept a careful food diary (not easy!)

over 24 hours or longer to see how their intake of the various components of their diet matches up with Recommended Dietary Allowances (RDAs).

Understanding the Nature of Science

At least as important as conceptual knowledge of nutrition is understanding the nature of science.[11] Such teaching should help learners appreciate issues to do with the reliability of nutrition guidance.

School students and adult members of the public generally see science either as being about certainty (as when it is used to plot the trajectory of a space rocket) or as being no more useful than any other set of views (as climate science is by those who reject the notion of anthropogenic climate change). One of the things we find it hardest to do in schools is to get students to appreciate that while much of scientific knowledge is held so robustly that in any useful sense of the word it is 'true' (e.g. that matter is made of atoms, that genetic information is conveyed from generation to generation though nucleic acids such as DNA and that entropy increases in a closed system), at the frontiers of science, knowledge is more provisional. This is why, for instance, scientists take time to develop reliable knowledge about such things as the extent to which humans are altering the climate, why some heterosexual couples who want to have children without the use of IVF find it difficult to do so and whether our solar system is utterly distinct or not that different from others.

Some aspects of nutrition come into the category of scientific knowledge that is held robustly — for example, there is no doubt that if your energy expenditure exceeds your energy intake day after day, you will lose weight (there is a limit to how much water you can drink). To illustrate the importance of taking both energy intake and expenditure into account, it isn't widely appreciated that the substantial increases in obesity rates in the UK over the last 40 years have been associated with an average decrease in energy intake. However, as we saw above with regards to fat and carbohydrate intake, while some important aspects of nutrition are well established, there is much that we do not know. The hope of better teaching about the nature of science is that people would realise that they should not throw the baby out with the bathwater. For example, the ongoing uncertainty about whether there might be health benefits from the occasional consumption of small amounts of alcoholic drinks should not blind us from accepting that large amounts of alcohol have major deleterious effects on our physical health as well as other disadvantages, being associated with a reduced capacity to exercise judgement and restraint, for instance when driving or when annoyed with someone.

Moral Education

Ethics is the branch of philosophy concerned with how we should decide what is morally wrong and what is morally right. We all have to make moral decisions daily on matters great or (more often) small about what is the right thing to do. Should I give money to animal charities or to medical charities? Is it fair that some people have plenty to eat while others go hungry? Should I give more weight to my interests than to those of others when choosing for whom to vote in an election?

We may give much thought, little thought or practically no thought at all to such questions. Ethics, though, is a specific discipline that tries to probe the reasoning behind our moral life, particularly by critically examining and analysing the thinking which is or could be used to justify our moral choices and actions in particular situations.[12] Food raises a number of moral and ethical issues,[13] including ones to do with labelling, safety, our use of animals, our use of land, inequalities between people and sustainability.

Traditionally, ethics has concentrated mainly upon actions that take place between people at much the same time. In recent decades, however, moral philosophy has widened its scope in two important ways, both of which are relevant to the issue of food: first, intergenerational issues are recognised as being important[14]; secondly, interspecific issues are increasingly being taken into account.[15]

Nowadays we are more aware of the possibility that our actions may affect not only those a long way away from us in space (e.g. the demand for cheap goods affecting the lives of workers in other countries) but also those a long way away from us in time (e.g. the use of nuclear weapons affecting generations to come). Human nature being what it is, it is all too easy to forget the interests of those a long way away from ourselves. Accordingly, a conscious effort needs to be made so that we think about the consequences of our actions not only for those alive today and living near us, about whom it is easiest to be most concerned.

Interspecific issues are of obvious importance when considering our use of animals and ecological questions. Put at its starkest, is it sufficient only to consider humans or do other species need also to be taken into account? Consider, for example, the use of new practices (such as the use of growth promoters or embryo transfer) to increase the productivity of farm animals. Many people feel that the effects of such new practices on the farm animals themselves need to be considered as at least part of the ethical equation before reaching a conclusion. This is not, of course, to accept that the interests of non-humans are equal to those of humans. While a small proportion of people do argue that this is the case, most accept that while non-humans have interests, these are generally less morally significant than those of humans.

Accepting that interspecific issues need to be considered leads one to ask 'How?'. A standard utilitarian answer is that we should consider just the pleasures and pains that would result from any action — but is this sufficient? For example, would it be right to produce, whether by conventional breeding or modern biotechnology, a breed of chicken unable to detect pain and unresponsive to other chickens? Such a breed would not be able to suffer and its use might well lead to significant productivity gains: it might, for example, be possible to keep it at even higher stocking densities than those currently employed. Someone arguing that such a course of action would be wrong would not be able to argue thus on the grounds of animal suffering. Other criteria would have to be invoked. It might, for instance, be argued that such a course of action would be disrespectful to chickens or that it would involve treating them only as means to human ends and not, even to a limited extent, as ends in themselves.

The Role of Food in Identity and Culture

A good education about food and nutrition would also help both school students and adults know something about the role of food in identity and culture. A number of health issues associated with diet have much to do with these factors. For example, there are dietary problems arising from the recent fad for so-called 'clean eating', endorsed by certain celebrities and wellbeing gurus. Precisely why some people pay more attention to the health pronouncements of such people rather than to doctors remains understudied but seems likely, at least in part, to be to do with how people see themselves in relation to others, which is a major part of what identity is all about.

Problems to do with 'clean eating' are recent (and thankfully relatively infrequent) but some eating disorders, for instance anorexia nervosa, are persistent, widespread and harmful. As is well known, anorexia (nervosa) is characterised by excessive food restriction, low weight and fear of gaining weight. The causes can be various and are often not well understood but are not infrequently psychological with a life-changing or stressful event quite often being the trigger. Physical complications include reduced fertility, heart damage and osteoporosis. In the West, about 1% to 4% of women, and 0.2% to 0.3% of men, will have anorexia at some point in their lives.[16] About 5% of those with anorexia die from it over a ten-year period.[17]

Teaching, both in schools and subsequently, that addresses eating disorders is important, given their prevalence and seriousness. Done well, it can help individuals to reflect on who they are, who they want to be and the forces around them that are trying to mould them. Part of me wants to write that this is particularly important for young women — and in many ways it is, given the gender-specific data on the incidence of eating disorders. But such education is important for young men too and it doesn't serve anyone's interests for them to be ignored or marginalised.

There are other, positive, reasons for considering the intersection of food and identity. To eat with someone can be an act of communion; nations can help define their own identities in part through the medium of food. The hackneyed phrase 'we are what we eat' may be largely incorrect to a dietician if understood narrowly — the extraordinary thing about nutrition is that pretty much whatever species one eats, one's body can change it not only into *Homo sapiens* but the particular human who is 'me' — but in cultural terms it has considerable truth. Jewish people over the centuries have chosen death rather than eat pork; and food that one does not want to eat can disgust, however nutritious it may be.

The importance of food to our culture and sense of self, as well as to our biology, suggests that as a topic it should be taught in a number of school subjects — Religious Education and possibly history and geography as well as biology. Many countries, including England, no longer teach 'domestic science' (aka 'home economics') in schools as they once did. In England, Northern Ireland and Wales, the introduction of the National Curriculum in 1989 washed away the subject, along with others like metalwork and woodwork, and replaced them with Design & Technology (D&T). There is much that is good about today's D&T, but the loss of good quality domestic science is a loss. Many young people do not learn about food preparation and cooking at home. Domestic science gave them that opportunity at school and also help counter the overemphasis in schooling on academic subjects.

Much of the above material on food and nutrition education has concentrated on school-aged learners. Some of the most interesting work on education about food with adults has been undertaken using participatory methods. In this approach, people whose expertise comes from their life experience rather than from professional training are seen as 'everyday experts'. In one project, work on people's indigenous understandings of food:

explores how these experts-by-experience can work together
with professionals to transform our damaged food system towards
health and agroecological goals on the one hand, and a fair
distribution of power, risk and resources on the other. We have
been brought together by a common desire to reflect critically
on how people's knowledge and wisdom can be harnessed
through action, participatory research and critical learning in
support of movements for greater justice in the food system.[18]

To give just one example, the Potato Park opened in Peru in 2001. It
is managed by Quechua communities who grow more than 1,400 potato
varieties in an important centre of crop diversity, thereby sustaining the
socio-cultural systems that have created and preserved this biodiversity.
Participatory action research has systematically affirmed Indigenous people's biocultural rights and has contributed to sustaining the capacity of
agriculture and food systems to adapt to change by actively guiding crop
evolution in the Potato Park. In addition:

Quechua cosmovisions have informed the development of new
concepts around biocultural heritage and solidarity economy
based on reciprocity. Rural women's organisations now manage
a polycentric network of barter markets that is important for
local food sovereignty and nutrition ... Quechua communities
have also successfully negotiated with the International Potato
Centre for a repatriation agreement on the return of more than
450 of their traditional potato varieties. These are now used in
farmer-led participatory research on climate change adaptation
(info.ippca.org), with local communities developing innovations to defend their Andean crops and indigenous knowledge
against biopiracy.[19]

Good education about food and nutrition can thus be emancipatory
as well as enable public understandings. Furthermore, such examples
illustrate the importance of context. While there are some basic principles
of nutrition education that apply broadly, good food depends on cultural
factors and so education about food should not be narrowly reductionist.

Food can play an important role in human flourishing and one wants this appreciated by people. Education about food and nutrition should, as argued above, include material on the nature of science, moral philosophy and cultural studies as well as biochemistry and digestion.

References

1. Wrangham R. (2013) The evolution of human nutrition. *Curr Biol* **23**(9): PR354–R355.
2. Cordain L. (2011) *The Paleo Diet: Lose Weight and Get Healthy by Eating the Foods you were Designed to Eat, revised edn.* John Wiley, Hoboken NJ.
3. Abrahams I and Reiss M. (2012) Evolution. In: Jarvis P with Watts M (eds.) *The Routledge International Handbook of Learning*, pp. 411–418. Routledge, Abingdon.
4. Leonard WR and Robertson ML. (1994) Evolutionary perspectives on human nutrition: The influence of brain and body size on diet and metabolism. *Am J Hum Biol* **6**: 77–88.
5. Rees M. (2018) *On the Future Prospects for Humanity*, pp. 171–172. Princeton University Press, Princeton NJ.
6. Ludwig DS, Willett WC, Volek JS and Neuhouser ML. (2018) Dietary fat: From foe to friend? *Science* **362**(6416): 764–770.
7. Seidelmann SB, Claggett B, Cheng S, *et al.* (2018) Dietary carbohydrate intake and mortality: A prospective cohort study and meta-analysis. *Lancet Public Health* http://dx.doi.org/10.1016/ S2468–2667(18)30135-X.
8. FAO. (2018) *The State of World Fisheries and Aquaculture 2018 — Meeting the Sustainable Development Goals.* Rome. Licence: CC BY-NC-SA 3.0 IGO. Available at http://www.fao.org/3/I9540EN/i9540en.pdf.
9. Byrne J and Grace M. (2018) Health and disease. In Kampourakis Kand Reiss MJ (eds.) *Teaching Biology in Schools: Global Research, Issues, and Trends*, pp. 74–86. Routledge, New York.
10. Scott C, Sutherland J and Taylor A. (2018) Affordability of the UK's Eatwell Guide. *The Food Foundation.* Available at https://foodfoundation.org.uk/ wp-content/uploads/2018/10/Affordability-of-the-Eatwell-Guide_Final_ Web-Version.pdf.
11. Reiss MJ. (2015) The nature of science. In Toplis R (ed.) *Learning to Teach Science in the Secondary School: A Companion to School Experience*, 4th edn, pp. 66–76. Routledge, London.

12. Reiss MJ. (2002) Introduction to ethics and bioethics. In: Bryant JA, Baggott-Lavelle LM and Searle JM (eds.) *Bioethics for Scientists*, pp. 3–17. Wiley Liss, New York.

13. See the *Food Ethics Council* website: https://www.foodethicscouncil.org.

14. Cooper DE and Palmer JA. (eds.) (1995) *Just Environments: Intergenerational, International and Interspecies Issues*. Routledge, London.

15. Rachels J. (1991) *Created from Animals: The Moral Implications of Darwinism*. Oxford University Press, Oxford.

16. Smink FR, van Hoeken D and Hoek HW. (2012) Epidemiology of eating disorders: Incidence, prevalence and mortality rates. *Curr Psychiatr Rep* **14**(4): 406–414.

17. Espie J and Eisler I. (2015) Focus on anorexia nervosa: Modern psychological treatment and guidelines for the adolescent patient. *Adolesc Health Med Ther* **6**: 9–16.

18. People's Knowledge Editorial Collective (eds.) (2017) *Everyday Experts: How people's Knowledge can Transform the Food System*. Coventry University, Coventry, p. xix. Available at: https://www.coventry.ac.uk/research/areas-of-research/agroecology-water-resilience/our-publications/everyday-experts-how-peoples-knowledge-can-transform-the-food-system/.

19. Pimbert MP, Satheesh PV, Argumedo A and Farvar TM. (2017) Participatory action research transforming local food systems in India, Iran and Peru. In: People's Knowledge Editorial Collective (eds.) *Everyday Experts: How People's Knowledge can Transform the Food System*, pp. 99–118. Coventry University, Coventry, p. 110. Available at: https://www.coventry.ac.uk/research/areas-of-research/agroecology-water-resilience/our-publications/everyday-experts-how-peoples-knowledge-can-transform-the-food-system/.

Part Three

The Social Context

11 Religious Implications of Multilevel Systems Biology

Niels Henrik Gregersen*

This chapter deals with the religious reception of Darwinism in the public space, and with the ways in which theological reflection can help bring versions of evolutionary theory to bear within religious self-understanding. One issue is the public perception of biology by religious onlookers, another issue is to absorb biological insights into theological frameworks.

The first section of this chapter points to the demographic fact that religions constitute an important cultural environment for evolutionary biology. I argue that since a considerable majority of the world's population is religiously affiliated, it is in the objective interest of evolutionary biologists to make friends rather than foes out of religious communities, wherever possible. Religious receptions of Darwinism are very diverse, however, some being hospitable to evolutionary thinking, others adversarial. Against this background, the chapter charts three models for relating biology and religion: a combative model of conflict, a model with a practical working division, and a more demanding model for possible interactions in the

*Professor of Systematic Theology, University of Copenhagen, Faculty of Theology, Department of Systematic Theology, Karen Blixens Plads 16, 2300 Copenhagen S, DK - Denmark.

public dissemination of evolutionary biology. Such a three-fold scheme is known from the general field of science and religion, but I will particularly emphasise that elements from all three models may be helpful in framing an evolutionary theology. More often than not, elements from several models can be brought into combined arguments. While controversy is a necessary strategy in some cases, the models of working division and creative interaction have a wider scope, and are more attentive to differences and details.

In the final sections of the chapter, I discuss the religious implications of evolutionary biology — from strict neo-Darwinism to systems biology — in terms of a set of 'what-if' arguments. First, *if* one believes in a creative and loving God, and *if* the general macroevolutionary account of evolution is overall correct — how can this be reflected in a religious understanding of life processes? Secondly, *what if* mutations and selection at the gene level are the only drivers of biological evolution, the result being survival or elimination? Thirdly, *what if* selection takes place at several levels (genetic, organismic, group, and species), so that selection includes creativity and retention (not only elimination), and selection takes several forms (some competitive, some cooperative)? I argue that while both a gene-centred neo-Darwinism and more complex versions of evolutionary theory have religious potentials, the systemic move of current biology opens up a new cognitive space for developing religiously more resonant views of evolution in a biological world that is holistic, multi-level, fluid, and creative.

Evolutionary Biology in Demographic and Cultural Perspective

According to a worldwide poll, about 84% of the world's population defined themselves as religiously affiliated in 2015.[1] Islam, Christianity, and Hinduism, in particular, are growing religions, while the group of unaffiliated 'nones' comprised 16% of the world's population in 2015. Given this background, it would be in the interests of evolutionary biologists to cultivate positive relationships with the world of religions, wherever possible. Doing so, however, one has to accept the fact that religions differ in kind, not only among themselves but also internally, also in their will and capacity to integrate findings and insights of evolutionary biology.

Demography matters but so does culture and religion. In a study from 2005 of the public acceptance of evolution in 34 countries,[2] Iceland, Denmark, Sweden, France, and Japan topped the list regarding the public acceptance of evolution with 83% to 78% adherents. In Denmark, for example, 82% accepted evolutionary theory in 2005, and in the same year 82% of the population happened to be members of the Evangelical–Lutheran Church in Denmark. Evolution and religion are not an either–or but more often a both–and. At the bottom of the European list lies Turkey with around 28% acceptance of evolution, and in the United States evolution had an acceptance rate of only 39%.

A more recent Gallup poll on the US population from 2017 is more encouraging for biologists as well as for evolutionary theists.[3] To the question: 'Human beings developed over millions of years but God had no part in this process', 15% responded positively in 2012, rising to 19% in 2017; this 'non-interference' group comprises atheists and agnostics but also people with a non-theistic spirituality, or groups of theists who are non-interventionists. The view that 'Man developed with God guiding' was affirmed by 32% in 2012, rising to 38% in 2017; this is the group usually termed as evolutionary theists. Finally, we have the group of creationists in the United States. In 2012 as many as 46% shared the view that 'God created humans in present form within the last 10,000 years'; in 2017 the figure had reduced to 38%. The two groups that accept evolution with or without divine influence comprise 57% of the US population in 2017, compared with 39% in the previous Gallup survey from 2005. Even if these figures may not be precisely comparable due to differences in methodology, there seems to be an overall trend in favour of acceptance of evolution.

From Quietism to Conflict Models

What are we to make of such demographic surveys of the public reception of evolutionary theory? One possibility is to take an attitude of wait-and-see. Such a quietist attitude may be natural for hard-working biologists who are more interested in exploring the biochemical mechanisms of life than in ever changing polls. To biologists concerned with the public understanding of biology, however, this is hardly a satisfying approach.

Another possibility is to take seriously the religious reception of evolutionary biology as a core component of the public reception of biological thinking. The interest in the public dissemination of biology, and of the religious resonance between biology and religion in particular, can take different forms, however. Based on the (statistically false) assumption that evolutionary biology and religion are born enemies, one option is to engage in an infight between Darwinism *in toto* and religions of whatever kind. This is a position well-known from atheistic evangelists such as Richard Dawkins but also from creationist evangelicals who blame Darwinism for all moral evils in society, while even questioning the validity of macroevolution based on fossils, carbon-14 methods etc. Indeed, it is impossible to avoid a conflict between a particular interpretation of a universalised Darwinism and religious positions that take the Bible to be the final arbiter of truth concerning natural history. Such combative approaches, however, make two problematic assumptions. One is that evolutionary theory and religion are fixed and unaltered positions; but just as evolutionary theory is itself evolving, and has been so from its early beginnings,[4] so religions are in constant movement due to ever-new self-interpretations. Another problematic assumption is that biology and religion offer rival explanations of reality at the same levels of reality.

However, logic defies the notion that 'natural selection' and 'God' can be rival explanations at the same level, as presented by Dawkins.[5] After all, natural selection takes place in populations of living organisms emerging from a physical cosmos hospitable for life, and biological processes remain sustained by underlying physical and chemical conditions. At the basic level of matter, it is physics that explains biology, and not the other way around. Similarly, for reasons of logical scope, the idea of God is not translatable into a particular scientific or counter-scientific hypothesis of empirical phenomena. If one were to see God as an explanatory hypothesis, it would be more like a metaphysical hypothesis, referring to the creative source of all that is, including the physical conditions for life, and the phenomenal outcomes of physical–chemical and biological processes. Accordingly, a rivalry between science and religion might better, if at all, be cast in terms of a rivalry between a metaphysical interpreted Materialism ('reality is nothing but atoms, molecules etc.') and a comprehensive view of God as the circumambient Reality: 'God is the eternal source of mass, energy information in with, and under material fields and elements'.

However, the meaning of religious God talk mostly relates to life as experienced within a more mundane context. In attitudes of basic trust, God may be perceived as a benevolent background condition, creating and sustaining the *framework* of the universe at its ultimate level (deeper than the level of biological explanations). First and foremost, however, God is perceived as present *within the boundaries* of human life experiences. It is within life experiences of excitement or hardship that religious people approach God in personal terms — in expressions of lament or praise, in prayer and meditation, in gratitude and hope, in ritual and ethical behaviour.

Biology and Academic Theology: Two *Post Hoc* Disciplines

Obviously, first-order lived religion is not concerned about offering alternative chemical explanations of evolutionary development, nor do religious reflections often relate to a multilevel systems view of the world of the living. However, just as evolutionary biology provides a *post hoc* empirical–theoretical explanation of the phenomenon of life at all its levels and all its manifestations, so academic theology provides a *post hoc* interpretation of the experiential concerns of lived religion. Without Christianity, for example, no Christian theology; without Islam, no Islamic theology, etc. While religious life is the search for ways of truthful living, theology is the second-order exploration of the internal reasonableness of lived religion, concerned also with the rationality of religious beliefs in relation to other beliefs that we hold true. For there exists no 'pure Christianity', for example, but only culturally embedded Christian churches, communities, and Christian ways of believing. Accordingly, just as the religious reception of biology belongs to the environment of evolutionary biology, so does evolutionary biology belong to the cultural environment of contemporary religion. However, it is the particular task of academic theology to reflect on the relations between religious and biological candidates of truth via interdisciplinary work.

To use an old distinction, biology is about offering a variety of context-dependent explanations of *how* nature works, while religious reflection asks the questions, *why* the world is as it is, and *why* it is as beautiful but also as tough, as it is experienced from a human perspective. These why questions are second-order reflections of religious life forms, subsequently explored in academic disciplines such as philosophy of religion and systematic theology. More often, however, lived religion is related to daily life concerns, and hence is relatively unconcerned about metaphysical explanations. Religious forms of life are therefore rarely concerned about backwards explanations but more often concerned about future-oriented issues, such as *where* to go and *what* to do in given situations. At this level, ethics is closer to religious life than science. (Observe, however, that already at the level of lived religion, religious people cope with issues well-known from evolutionary biology, such as finding a balance between competition and cooperation, and aiming to cultivate human lives in the face of the inescapable dilemmas, including the biological dilemma that we always live and survive at the expense of some other life.)

The Pragmatic Model of Working Division

Unlike the combative position, the model of working division is attentive to the differences of epistemic interest between evolutionary biology and lived religion. Famously, the Harvard biologist Stephen Jay Gould coined the principle of NOMA (Non-Overlapping Magisteria) between religion and science, well captured in the quip that one thing is the Rock of Ages, another thing the age of rocks. The first belongs to the competence of theology, the latter to the competence of science.[6]

In my view, Gould and his followers are right in reminding us of the practical working division between science and religion, and between biologists and theologians. The NOMA principle has the advantage of freeing religiously disinterested biologists from being overly involved with religious friends and foes of evolutionary thinking; it also has the advantage of freeing biologically disinterested religious people from the misunderstanding that religion stands and falls with subscribing to a particular view within evolutionary theory. There are indeed quite some gaps of epistemic interest between evolutionary theory and religious views of life. One way of restating this difference is to say that while the pursuit of science is typically specialised, and thus has an endemic *narrow perspective* as the precondition for scientific precision, religious views of life are *broad-scale* in perspective, but for the same reason much less precise. At the same time, religious perceptions of life are usually *limited* to the experienced phenomena of life, whereas biology explores the underlying biochemical set-up and dynamics of cells and organisms as part of the explanation of the phenotypical world of cells, organisms, populations etc.

Due to the logical scope of beliefs in God as the all-encompassing reality, religious commitments cannot rely on particular versions of evolutionary theory. The Harvard biologist, Martin A. Nowak, offers an example of such a relaxed view of the relation between biology and religion. Nowak understands evolution as exploring a 'space of possibilities'[7] (p. 50), subsequently to be investigated by different biological methods, including molecular biochemistry and mathematical game theory. Being convinced that the initial transition from chemistry to the first instances of life will one day be fully explained (which it is not today), he is strongly committed to closing any remaining explanatory gaps within science by way of future scientific progress. Accordingly, his religious commitment does not rely on any god-of-the-gaps argument. Rather, in the vein of St Augustine and St Thomas, he understands the creator and sustainer of the universe to be an atemporal God, for whom 'the evolutionary trajectory is not unpredictable but fully known'[7] (p. 51). This is an example of 'high' theology coupled with a clear working division between biology and theology. Science deals with the *causal exploration* of the realisation and potentialities of evolution, while theology and philosophy deal with the issue of *metaphysical explanation*, relating to all that is, including the evolutionary phase space.

Difference-Based Interaction: Life as a Shared Concern

A good sense of the working division between science and religion should not, however, block us from acknowledging the shared concerns between biology, on the one side, and theology and philosophy of life, on the other. We here arrive at the third and potentially more interactive model for the relation between religion and evolutionary biology.

Beneath the territory of distinctive biological and theological theories, there is a common concern about the phenomena of life. 'Life' is prior to both biology and theology, and it is not the case that the disciplines of biology tell us what life 'really' is, whereas experienced life processes are not as real as the underlying biochemistry of life. Rather, everyday human experiences of life constitute the raw material for religious and philosophical reflection, and life experiences, such as birth and death, joy and pain, competition and cooperation, are so important features of life that they are addressed in ordinary life situations as well as by biological reflection.

There may be a long way from the labs and writing desks of working biologists to human lives and religious modes of existence, but the *flourishing of life* is a core value for human beings, regardless of worldview differences. More than that, not only humans but also other mammals and many other species take care for the survival of their offspring, just as human beings are usually concerned about health issues and future perspectives of life. The generation of care and self-care seems to be a general result of the complexity of evolution, involving a transfer of information from the chemical level to the level of skilfully exchanged body signals, in the case of humans also though the symbolic exchange of language and resulting cultural patterns of passion and compassion.[8]

The same goes for the darker sides of existence such as pain, anguish, death. Also here, one finds correlative insights amidst differences of perspectives. Philosopher of biology Michael Ruse (himself a religious sceptic) has pointed to the obvious fact that '[b]eing a Darwinian does not compel one to be a Christian'[9] (p. 142). He adds, however, that 'if you are a Darwinian looking for religious meaning, then Christianity is a religion that speaks to you. Darwinism, a science which so stresses physical suffering, looks to Christianity, which so stresses physical suffering and the divine urge to master it'[9] (p. 131). Drawing on the history of biological thought — from Darwin's contemporary Asa Gray to champions of the Modern Synthesis such as Ronald Fischer and Theodosius Dobzhansky — Ruse points to the fact that leading proponents of evolutionary theory were Christians who themselves emphasised the convergences between Darwinism and central Christian beliefs. Ruse's general argument may be seen as a gentle gesture to religious people from an astute Darwinian, but Ruse's argument may also be read as an attempt of a damage control with respect to the anti-evolutionary views of Protestant fundamentalism, distinctive to US American religious culture. In a similar way, a sociobiological hardliner such as Edward O. Wilson published *The Creation: An Appeal to Save Life on Earth*[10] in which he calls for an alliance between concerned biologists and religious people concerned about the future of life on planet Earth. Both share the commitment for the continuous flourishing of life.

Coming from the religious side, some even speak of the 'biologization'[11] (p. 69) of contemporary worldview conceptions, at least in the minimal sense that the question of meaning is tied to concepts of the flourishing of life, hence formulated within a largely immanent framework. The Swedish philosopher of religion, Carl Reinhold Bråkenhielm, has worked on the question of the biologisation of worldviews in contemporary Sweden, and has pointed to the curious fact that the statement 'Biology gives us the complete answer to the meaning of life' is affirmed by proportionally significantly fewer Swedish academics (13%) than by the Swedish population as a whole (24%). On the other hand, the statement 'Nature is filled by a spiritual force permeating all life' is affirmed by 42% of the Swedish population and by 25% of Swedish academics[11] (p. 70). The responses seem to indicate that 'life' is perceived as something bigger than what is analysable in terms of the biological sciences. What seems to be the case is that the findings and results of evolutionary theory, while generally accepted, are *received* as part of a bigger package of insights into what life is all about, what it means to live, and where the meaning of life may be found from an engaged, first-hand perspective.

Some General Religious Implications of Evolutionary Theory

Given that there are not only differences of methods and interest between evolutionary theory and religious modes of life, but also substantial overlaps of concern, I'm now going to entertain a small set of thought experiments in terms of *what if* questions. Accordingly, I now change the perspective from discussing *public receptions* of Darwinism — seeing evolutionary theory in its cultural contexts — to theological reflections on the *religious implications* of different versions of evolutionary theory. How is a religious self-reflection, mediated by theological thinking, able to build central biological insights into its own wider understanding of what 'life' is all about?

   Observe that 'religious implications of biology' is here not taken in a strict logically sense, as if evolutionary theory implies a very definite shape of belief, or a lack thereof. One can be an evolutionary Darwinist, and yet be a Christian, Muslim, Hindu, Buddhist, an Orthodox Jew, an agnostic, or an atheist. Likewise, religious convictions are built on a variety of resources such as religious texts, songs, ritual experiences, and everyday experiences, and only rarely does lived religion base itself on particular biological theories. Theology, however, should be more interested than are religious practitioners in evolutionary theory-making. For, as we saw above, theology is not confined to the study of the inner logic and meaning of religious modes of life within given religious traditions, but is also concerned with the rationality of religious beliefs in relation to other beliefs that we hold to be true. Accordingly, we should acknowledge that 'theological implications' appear only where theologians actively bring the potential messages of evolutionary theory to bear within theology itself.

We here face a certain asymmetry between biology and theology. In the pursuit of evolutionary biology, no religious interpretation is required in the labs, and no reference to God should be used within biological theory. By contrast, academic theology has to be interdisciplinary in nature, since it deals with the relation between religious truth candidates and the strongest truth candidates available, including those of evolutionary thinking.

Some of the more general take-home messages from Darwinism have already long been absorbed into mainstream Christian theology. The contours of what might be broadly called Evolutionary Theology[12] was already formulated in Darwin's time, beginning with Darwin's own *On the Origin of Species* from 1859, his autobiography, and letter correspondence, but more fully developed forms of evolutionary theology only emerged during the 20th century. We here find extensive thought experiments worked out in detail: What happens if you're a convinced Christian, and if you find the basic Darwinian arguments convincing as well? Three issues have been at the forefront: Biblical interpretation, human uniqueness, and the problem of evil.

Long before Darwin, the view that the Bible is a handbook in natural history had been eroded by the 'higher criticism' of Biblical interpretation, beginning by the end of the 1700s in Germany. Based on Charles Lyell's *Principles of Geology* (1830–1833), Robert Chambers' *Vestiges of the Natural History of Creation* (1844) argued that God works through the uniformitarian laws of nature. Accordingly, theologians began to interpret *Genesis* chapters 1–11 as a foundational myth about the perennial human condition rather than as an account of discrete epochs in early natural history. Precursors for this spiritual interpretation can already be found in the Jewish philosopher-theologian Philo of Alexandria (c. 20 BCE–CE 50), and in 4th century church fathers such as Gregory of Nyssa and Augustine. Today, theologians often refer to the 'common creation story' provided by the Big Bang cosmology and evolution, and religious 'metanarratives' that focus on the question of the meaning of selfhood, social life, and relations to divinity.

A second and more difficult question has been the role of humanity in an evolving cosmos. To many of his contemporaries, Darwinism was simply identical with the 'ape-theory', and still today common ancestry is a disputed feature, particularly in the Muslim world.[13] Proponents of evolutionary theology have developed several arguments against the view that the Darwinian view of common descent implies a humiliation of the human race. The principal argument (already developed by 19th century Kantians) has been to apply the fact–value distinction, arguing that the dignity of humanity does not depend on its separate historical origins but concerns the intrinsic value (or disvalue) of humanity as it is experienced today. Without any prehistorical Adam and Eve, there is still a distinctiveness of human beings,[14] intrinsically related to the observed trends of complexification during evolution.

The Darwinian notion of common descent, however, has not only served as a detraction from the earlier view of the separate creation of Adam and Eve but also as an inspiration for exploring the relations between humans and other co-creatures. Evolutionary theory has pointed to the continuity between human beings and all other life forms, and to important analogies with higher animals. Central aspects of human cognition find precursors and analogies in other species (such as brain development), locomotion (including but not confined to bipedal walk) and the development of sentience (from light-sensitive cells to compound eyes and lens-based eyes). Likewise, in social insects we find systems of division of work; among higher apes we find forms of behaviour that may be termed proto-moral. This comparative perspective even involves cases in which animal development far exceeds that of humanity, for example, in terms of physical strength, the olfactory senses of dogs, and the evolution of forms of awareness unknown to humans (such as echolocation in bats and dolphins). Without humiliating the human species, such evolutionary features have indeed elicited a sense of humility on the part of humanity. Accordingly, evolutionary theologians see humanity as being peers with non-human creatures, and influential theologians such as Jürgen Moltmann have argued that both evolutionary and biblical perspectives require us to understand humanity as created in the image of the world (*imago mundi*), before it makes sense to speak of humanity as created in the image of God (*imago Dei*).

The third and hardest problem for an evolutionary theology is the problem of pain, suffering, and death. However, one should not see Darwin as the one who has placed the onus of the problem of evil on religious thinking, for this problem is from ancient times reflected in the Book of *Job* and many other religious writings. In two ways, however, contemporary theology has to work with new challenges compared with pre-Darwinian theology. First, interpreting the story of Adam and Eve as a myth of the perennial human condition undermines the traditional view that a prehistorical Fall explains how and why evil and death came into the world of the living. It is no longer human sinners but the creator of the universe who is ultimately responsible for the construction of a world with non-human pain and death, and with human suffering and anxiety. Secondly, the Darwinian theory shows how and why pain and death are systematically built into life processes, so that the world of creation has to be acknowledged as a package deal of joy and woe. As aptly expressed by the German biologist Rupert Riedl, 'It was first with multicellularity that death came into the world, first with the nervous system that pain came into the world, and first with consciousness that anxiety came into the world'[15] (p. 290). Hence, evolutionary theologians have to accept biological death as natural (not as a guilt-ridden curse), and understand that death, pain and mental suffering are the price to be paid for living in a developing world with highly complex and intense forms of sentient life. The only alternative is the *anaemia* of living in a trivial world, without sentience and therefore also without the subjective experience of suffering and decay.

Evolutionary theologians, however, are divided on the issue whether biological death has the final say about the future of human lives, or whether there is also a future redemption possible from the world of pain and death. Some Christian evolutionists argue, for example, that God is not only the creator of an ambiguous world of creation, but is also the co-sufferer who carries the costs of complexification within divinity itself. In the story of Jesus, seen as the 'Son of God', he takes not only human anxieties but also the experiences of painful co-creatures into divine experience in order to refashion new life in an all-encompassing divine reality beyond the immanent framework of biological lifespans.[16,17] The example shows that a Christian evolutionary theology may utilise the resources of all three articles of the Christian faith, referring to God as a fatherly creator (1st article), to God as incarnate into the material depths of creation in the life story of Jesus the Christ (2nd article), and as the redeeming and vivifying divine Spirit (3rd article). In so doing, theology transcends the biological perspective, with the unavoidable consequence that there is here no longer a direct traction between theology and biological theory. However, basing religious commitments entirely on evolutionary theory presupposes that evolutionary biology is not only complete in itself but also that biology is omnicompetent, providing the correct answers to all kinds of metaphysical and theological issues. Such an exotic view is both scientifically and philosophically untenable, and it also serves as a reminder that a theology operating solely at the mercy of Darwinism will inevitably end up as a merciless theology that leaves behind evolutionary losers on the bloody battlefields of evolution.

What If Neo-Darwinism were Sufficient to Explain Evolution?

The thought experiment to be pursued now asks what might follow for a religious view of reality if a strict neo-Darwinism were true. By 'strict neo-Darwinism', I here refer to a specific development within Darwinism described by David Depew[18] in this volume as a gene-centred adaptationism.

The so-called Central Dogma of molecular biology, first formulated by Francis Crick in 1957, states that DNA acts as the template for making RNA, which in turns acts as the template for making proteins. This gives the picture of a seamless pathway: DNA → RNA → proteins. The identification of the double helix structure of the genome by Francis Crick and James Watson in 1953 paved the way for the subsequent success of the gene-centred view of evolution. Hereby, Charles Darwin's broad principle of variation, which included changes in the environment, was itemised as *mutations*, that is, aberrant changes that by chance take place at gene level.

In strict neo-Darwinism, the biochemical structures of genes are viewed as the primary coding engines of cells by producing single-strand RNA etc. In a later reformulation, however, Crick wisely restated the Dogma in a more cautious form: 'The central dogma of molecular biology deals with the detailed residue-by-residue transfer of sequential information. It states that such information cannot be transferred back from protein to either protein or nucleic acid'[19] (p. 561). Hereby, Crick avoids the problem that while DNA regularly codes for RNA, there are also important cases in which the RNA codes backwards on the DNA, so that we in the end find examples of a circular network: DNA → ← RNA → proteins. Moreover, RNA is often able to replicate itself into other RNA without the help of DNA (as in many viruses). With these modifications, the Dogma is wrong when taken as a unicausal pathway, but still works for most practical purposes.

A second reason for singling out the genome, according to strict neo-Darwinism, is that the genes are the only carriers of information to the next generation. The reproduction of organisms, however, takes place at the level of organisms in their ever-shifting environments. Here the *mechanism of selection* is at work at two distinctive levels. It is the phenotype (the individual organism) that is the *unit of selection* that is either eliminated before reproduction, or succeeds to reproduce itself before its decay. However, it is only the genotype (DNA) that is the *unit of reproduction* to be transmitted into the next generation, where the DNA → RNA → protein route starts over again. Hence, the differential capacity to reproduce genetic material is what ultimately matters in evolution.

Thus, selection processes provide the net result of the reproductive success of a gene constellation, compared to the gene constellations of other organisms within the same population. This is what is often referred to as 'the survival of the fittest'. But adaptive fitness is finally about 'fitting in' to the extra-organismic environmental conditions. Adaptation is not about having the best genes but about having the most suitable genes in a given environment.

The end result of the route leading from genes to reproduction (via the vehicles of the organismic phenotypes) is their reproductive success. Sometimes, evolutionary biologists express the distinction between genetic mechanisms and the measure of their reproductive success in the distinction between proximate and ultimate causes. 'Proximate causes have to do with the decoding of the program of the given individual; evolutionary causes have to do with the changes of genetic programs over time, and with the reasons for these changes'[20] (p. 68). *Proximate causes*, in other words, are the genetic mechanisms at work within the organism, whereas *ultimate causes* are the evolutionary forces at work over generations, depending on the differential success of reproduction and change of organisms — the net result of selection.

I hope to have given a fair (but certainly not complete) description of the gene-centred adaptationist program of neo-Darwinism. I happen to have worked theologically on the premises of neo-Darwinism when I was young and not yet aware of broader programs within Darwinian biology.[21] Then, as now, I believed that a theological response congenial to neo-Darwinism should avoid any general criticism of the 'reductionism' involved in the program. Reductionism is a scientific method to be welcomed wherever it works, just as conceptual analysis is a helpful method in theology and philosophy as long as it takes us.

After all, the gene-centred approach has been extremely successful precisely due to its reductionist methods. One example is the Human

Genome Project which was formally launched 1990 and declared complete in 2003, after having mapped and sequenced the nucleotide base pairs that make up human DNA — with the unexpected result that the human genome consists of only about 20,000 genes, most of which we share with other animals. More recently, in 2007 and 2014, the first two phases of a corresponding Human Microbiome Program were launched by the NIH. Samples have been taken out from 15 to 18 human body sites — from the gut and nostrils to skin — and systematically sequentialised as the non-human DNA and ribosomal RNA living in human bodies. Like the Human Genome Project, the Human Microbiome Project is likely to be important for future medical treatment. But the project also shows how much non-human DNA and RNA we carry within us, including bacteria, eukaryotes, archaea and viruses; bacteria alone outnumber human cells by an order of magnitude.[22] Biologically speaking, we are more than ourselves, and what we are and can do as human beings depends on the microbial communities living within and on us. We better have to make friends with other organisms dispersed in our bodies, for most of them are our helpers. This example shows that reductionist methods can have results that are of central importance also for a theological anthropology that takes interest in the evolution of symbiotic alliances and community building over time.

The need for supplementing reductionist methods with more holistic viewpoints in biology arises since we otherwise lose sight of the wider networks of interacting causes in the world of biology; this is similar to the way that well-winnowed analytical approaches in philosophy and theology risk losing sight of important phenomena that arise in the connections between itemised themes. Let me add that I find it misplaced if theologians wish to bring in God as an additional causal factor in the explanation of biological features. God is not a foreign interventionist to be added to creaturely life processes. If God Is, God is both the transcendent source of all material structures and immanently present in, between, and under creaturely processes. For the same reason, the term 'ultimate causation' refers in theology to the divine grounding and loving sustenance of all creaturely creativity at the deepest metaphysical level; it is not to be identified with the successful net result of evolving life, as in the evolutionary concept of 'ultimate causation'. God is the ultimate creator of the whole texture of the material world as a biofriendly universe, but religiously it is likewise important to understand God as immanently present in, with, and under all processes of life — both in its splendours and in its processes of decay. As infinite, God cannot logically be limited to what is beyond the world, but must be the circumambient Reality hosting life processes within Godself (a view sometimes referred to as panentheism).

Any such theological view, however, will gain in plausibility if the biological world produces qualities that carry at least some similarity to what Christians and other theists call divine fecundity and generosity. A counterexample may be the appeal to mutations as evidence of a cosmic chance, devoid of any meaning. In his well-written and widely read book, Chance and Necessity,[23] the French biologist Francois Monod made the famous picture of human beings living as gypsies at the edge of the universe, desperately making meaning out of a pointless and adversarial universe. The Oxford biologist Arthur Peacocke, however, has rightly seen this interpretation as an example of making the good biological principle of variation into a problematic worldview argument: 'Unlike Monod, I see no reason why this randomness of molecular event in relation to biological consequence has to be raised to the level almost of a metaphysical principle in interpreting the cosmos ... It would be more accurate to say that the full gamut of possible forms of living matter could only be explored through the agency of the rapid and frequent randomisation which is possible at the molecular level of the DNA'[24] (p. 55). Indeed, mutations express the fecundity of the evolutionary possibility. In theological interpretation, God is not only the creator of a world with fixed parameters (such as laws of nature), but is also the source of potentiality and creaturely creativity.

Regarding the central evolutionary concept of reproduction, evolutionary theologians may also point to evolution as indicating a divine urge to call forth a biological world characterised by erotic desire, attraction, and sense of beauty. Strict neo-Darwinists have argued that falling in love is a mechanism fobbed upon us by evolution, but this metaphor of nature's cheating is both anthropomorphic and does not capture the flavour of life; in other words, Darwinian descriptions do not always 'save the phenomena' of lived life. The upwards search for sunlight, the outward search for nutrition, the song of birds, the calls of wolves, and the scents of lions may serve several purposes. One is to compete for resources, to protect territories but another is to call upon other birds, wolves, and lions. Sending out scents may well be described as a mating mechanism, a part of sexual selection, but eventually mating often leads to something like erotic desires, satisfaction, and the care for offspring. Again, sexual satisfaction may also be identified by higher levels of oxytocin etc. in animal brains, but we would not know about the importance of hormones and neurotransmitters if those signals were not correlates to mating behaviours and community building. In this sense, theology, together with real-life naturalists, may have a role in redescribing a biological world already described in more dire, neo-Darwinist language, and should do so in order to save the phenomena of life as experienced at the level of higher-order creatures. The expressions of life are always more rich and complex than in gene-centred theories thereof.

It is not an offence to state that neo-Darwinism does not provide a complete theory of organismic development and large-scale evolution. A complete biological theory would at least have to explain why macroevolution has led to a further complexification of the world of living systems. In an excellent analysis of inner-biological discussions, biologist Daniel W. McShea concludes that '[t]he case for increasing complexity in multicellular organisms is weak'[25] (p. 644). In a more recent publication, however, co-authored with the philosopher Robert Brandon, he argues that while natural selection explains adaptation, the trend towards complexification of evolution can't be explained by reference to Darwinian selection processes. They point a deeper-seated law of complexity, called the 'zero-force evolutionary law'.[26] In pointing to the inner limitation of the program of selectionist adaption theory, McShea and Brandon agree with the information theorist David Wolpert who refers to the 'depletion forces' of the second law of thermodynamics as leading to a robust growth of informational structures over time.[27] We here again arrive at the conclusion that biological processes are enveloped in more general physical processes that underlie and sustain them.

Speaking of a trend towards complexification in evolution does not imply that we are able to predict future processes of complexification. Living in an evolving biosphere, we probably have to acknowledge that there is not only a finite number of potential species. As argued by the complexity theorist Stuart Kauffman, 'evolution does not cause but enables future evolution'[28] (p. 181). On this basis it is reasonable that 'life' and the 'complexification of life' are features that require an exploration from different theoretical angles. Just as organisms are ever developing, so is speciation ongoing. In this sense, evolutionary biology is better equipped to offer retro-explanations than predictions. It is therefore unlikely that there will ever be something like a complete biological theory. It should be added, however, that religions are definitely not competent to fill the theoretical gaps and empirical *lacunae* of evolutionary biology. The model of working division is here in place. Theological reflection remains primarily in a learning position in relation to questions that fall under the competence of the physical and biological sciences. What it may offer is a wide-scope sense of coping with life experiences at the level of the phenomena.

What if the Multilevel Approach of Systems Biology Explains the Phenomena Better?

As shown by David Depew[18] in this volume, the history of 20th century Darwinism is a largely untold story of a rivalry between a more rigid, gene-centred adaptationism and the broader paradigm of developmentalism, in which evolution takes place at the level of organisms within populations that are capable of forming new species under certain environmental conditions.

While the Danish botanist Wilhelm Johannsen coined the term 'gene' in 1909, long before the Modern Synthesis of Darwinism and Mendelianism took place in the 1930s, the British embryologist C. H. Waddington coined the term 'epigenetics' in 1942 in the vein of the Modern Synthesis. His idea of the 'complex system of interactions underlying the epigenetic landscape'[29] (p. 64) is clearly framed within the developmental tradition of evolutionary biology, as presented in his visual models of evolutionary valleys (pointing to a low level of activity) in some parts of the cell, and of evolutionary hills in other sections of the cell (depicting a higher level of re-activation). However, neither Darwin and Mendel nor Johannsen and Waddington had any intimations of future development of computational techniques, by which we can handle large-scale information by sequencing and recombination. But as we saw in the previous section, bioinformatics was systematically employed in the Human Genome Projects as it is in the ongoing Human Microbiome project. Eventually, computational approaches of 'systems biology' are already in use across the distinction between more gene-centred and more environmentalist-oriented approaches.

In what follows I will not be able to discuss the quantitative possibility of bioinformatics (see Gatherer 2019[30]), but will discuss the vistas opened by multilevel systems biology. After all, the DNA–RNA circuit with all its active associates in the form of lipids, carbohydrates and protein foldings leads to a cellular development in which (once its structure is formed) much happens relatively unaided by the genetic setup. What is more important is the activation or suppression of genes, which relies on epigenetic stress factors, leading to genetic assimilation for specific purposes, such as the thickening of skin, or the response to overheating[29] (pp. 262–270). What biologists working in labs say is that it makes not much biological sense to speak of a genome plus an epigenome; rather, the genome itself becomes an epigenome by virtue of being enveloped in a wider cellular system that recodes, regulates, activates or deactivates a wide array of gene expressions.

As argued by Ottoline Leyser and Harris Wiseman[31] in this volume, systems biology must also proceed in *reductionist* mode in order to make complex systems tractable for empirical investigation: perturbating living organisms, isolating sequences of DNA and RNA, and identifying protein foldings is a precondition for being able somehow to monitor the interactive networks between DNA, RNA, proteins etc. This pragmatic approach of a *methodological reductionism* differs widely from the view of *ontological reductionism,* which would argue that the internal biochemistry of the double helix structure of the genome unilaterally explains the development of an organism. Contemporary biology seems a far cry away from the sociobiological announcements of the 1970s: 'the genes made me do it'!

Given that there is a synchronous system of feedback, it becomes difficult to make predictions without using computational models in tandem with wet experiments in the lab[32] (p. 517). What is taken as a given point of departure from one subset of investigation (say, DNA and RNA) is part of wider cellular whole, which again is part of a wider community. Genes operate together in clusters of genomes, chromosomes in eukaryotic cells, and cells cooperate in multicellular organisms, which again are parts of ecological networks.

In this sense, systems biology opens up the opportunity to analyse cross-over effects of many levels, from genes to the wider biochemical system of cells to multicellular organisms to organismic behaviours and to the interplay of species in wider ecosystems. Evolution, in this view, is not so much about exploration of the temporal linear evolution as connected to the exploration of the ecospace.[33] One example is the much discussed niche construction theory (see Laland[34] for critiques and counter-critiques). Here, the unit of reproduction is not only the genes but also the inheritance of abiotic structures (such as beaver dams) that are given over to the next generation of biological learners.

On the Outcomes of Evolution: The Case of Cooperation and Altruism

Let me focus on one case, in which many different levels come into play: the possibility of *cooperation*. Survival of organisms does not very often come about without cooperation, and examples abound even with regard to simpler lifeforms that have been with us since the early phases of evolution. One example is the demosponge *Amphimedon queenslandica* at the Great Barrier Reef, in which we find genetic mechanisms allowing cells to cooperate while suppressing individual cells that would destabilise the collective, if they were to multiply themselves extensively. Yet, in the context of the demosponge, a division of work takes place. As pointed out by Martin A. Nowak, such cooperation is not an exceptional case but is a recurrent feature of evolutionary history that offers a protective level of resilience for individual organisms in cooperative communities:

> In fact, coagulation of complex cells into cooperative communities was such a winning strategy that it evolved several times. Animals, land plants, fungi, and algae joined a communal life, and not just of their own kind. Coral reefs, the biggest living structure on Earth, are formed by an enduring partnership between an animal (polyp) and plant (algae), held in permanent embrace by a skeleton of limestone … Given how many times multicellularity evolved, it seems unlikely that there is a single explanation for its origins, save that the same basic strategy — cooperation — was the right answer when it came to dealing with various problems[35] (p. 140).

Any strategy of cooperation, however, has to measure up to tough biological requirements. One is the barrier that complex multicellular organisms are up against the stream of entropy, leading to the imminent danger of falling apart, when their environments do not provide enough energy and nutrients usable for the continuous reconstruction of the superorganism. Another biological wall is the fact that individual genes and cells can 'defect' and, so to speak, grow independently and fast at the expense of the collective organism. One lesson of mathematical game theory is that the dynamics of cooperation can be undermined: 'every stage of cooperation comes with the risk of defection. Cooperation is never stable.

There is a surge in cooperation during early development and childhood, but our cells begin to rebel with age' (Nowak 2011, 141). A well-known example of defectors are cancer cells that operate against the interest stage of the body of which the cancer cells are part. 'Cancer is the price we pay for having complex bodies built by an extraordinary level of cooperation'[35] (p. 141).

Using an anthropocentric metaphor, Richard Dawkins has spoken of 'selfish genes' from a strict neo-Darwinian perspective. In the meantime, other biologists point to the recurrent features of symbiosis and cooperation, what might similarly be designated as 'unselfish' genes, cells, and so on. From a philosophical point of view, however, one should treat popular references to selfishness and unselfishness with the utmost caution, reminding ourselves that they, popular as they may be, are anthropomorphic projections. Within a more robust biological reasoning, references of 'selfishness' or 'unselfishness' should be shunned, just like theologians should not fall into the temptation to speak of 'sinful' or 'saintly' genes.

The difference between egoism and selfishness in the human realm and natural selection operating at purely genetic and organismic levels is the presence of moral intentionality. Humans may put themselves, their group, or their country first intentionally, and even programmatically. Mechanistic operations of biochemistry, however, are devoid of intention, without other 'goals' than functionality. In between, we find examples of intentional behaviour, for example in the hunting behaviour of reptiles and mammals. Hunting practices are intentional, in terms of being clearly goal-directed; they are also often cooperative, yet without moral concerns towards their prey. The example shows that cooperation is a fundamentally amoral phenomenon. In their seminal work on group selection, *Unto Others: The Evolution and Psychology of Unselfish Behavior*, the biologists Elliott Sober and David Sloan Wilson state unreservedly: 'Group selection favours within-group niceness *and* between-group nastiness'[36] (p. 9). Their book is dedicated to 'Altruists everywhere, especially those who are unsure as to what their motives really are.'

In their essay, 'Why Cooperation Makes a Difference', theologian Sarah Coakley and biologist Nowak make it clear that we need to distinguish carefully between behaviours that involve *de facto* distributions of costs and benefits and behaviours that are intentionally held, and therefore open for moral characterisations. Accordingly, Coakley and Nowak avoid the standard term of 'biological altruism' and choose instead the term 'cooperation', which they define as follows:

> Cooperation is a form of working together in which one individual pays a cost (in terms of fitness, whether genetic or cultural), and another gains a benefit as a result.[37] (p. 4)

In his 'Five Rules for the Evolution of Cooperation'[38], Nowak presents a set of distinct game–theoretical models for different forms of cooperation based on the Prisoner's Dilemma. The first three versions are well-known from previous biological discussion, namely 'kin selection' (taking care of the holders of one's own genes), 'direct reciprocity' (also called tit for tat), and 'indirect reciprocity', in which the helpers gain a higher social reputation by being known as helpers. The two next forms of cooperation take place in broader social contexts. 'Network reciprocity' means forming clusters of agents who are prepared to help one another but not others, while 'group selection' takes place in communities, in which the group as a whole is likely to grow by the presence of a good portion of helpers, and with rules of punishment for defectors. However, as long as defectors go unnoticed, they have better chances of survival within the group than do the cooperative agents.

All these forms of cooperation are also known from human interactions: care of offspring, contracts of trade, striving for recognition and reputation, strong network building, and peer group mentalities. In the standard biological literature, all these forms of cooperation are termed 'altruistic'. Coakley and Nowak, however, propose a strong distinction between biological and social cooperation on the one hand and altruism on the other hand. Cooperation is only properly altruistic if it refers to a behaviour motivated by love, in which human persons intentionally seek benefits for others even at the cost of the altruistic individual:

> Altruism is a form of (costly) cooperation in which an individual is motivated by good will or love for another (or others).[37] (p. 5)

I take these clarifications to be important, both for clarifying what evolutionary biology does and does not teach about the many ways in which evolution works. However, it remains the case that altruistic persons can be in doubt about their ulterior motives, and motivations are difficult to validate empirically and measure accordingly.

Theological Perspectives

In his theological response to Coakley and Nowak's distinction between cooperation and altruism, Philip Clayton has proposed that 'a theological account might provide part of the explanation for acts of radical altruism'[39] (p. 349). His argument goes as follows: if reductionists are right, human mental states must be explained at the level of brain structures and bio-chemistry, ultimately down to the level of microphysics. This has ramifications also for theology: 'If beliefs about deity are sufficiently explained in purely empirical terms (i.e. by evolutionary biology, evolutionary psychology, cognitive science, etc.), it becomes arbitrary to appeal to the *actual* existence of God in order to account for why such beliefs exists'[39] (p. 353). The picture, according to Clayton, shifts the moment one realises that the opposite view of strong emergence is right: higher-order levels, such as human personhood, can't be explained from lower levels. Clayton is fully aware that the view of strong emergence does not in itself imply theological considerations. There is in Clayton no argument in the mode of a 'natural theology' that leads from strong emergence to the reality of the divine. However, Clayton constructs a similar argument in terms of a 'theology of nature', arguing that strong emergence is '*necessary* if theological assertions are to play any role in explaining any human behaviors'[39] (p. 353).

I fully agree with Clayton that theology works better under conditions of strong emergence than on a unilateral causation from below. Nonetheless, I find the hypothesis that the reality of God may 'causally' explain particular features of evolution too *ad hoc*, and too external, as if God is brought in as a particular causal explanation of radical altruism in human lives but has nothing to do with other forms of cooperation, for which we already have other (apparently sufficient) causal explanation.

My own theological proposal is to take a more integrative view of religious truth claims while also wanting to acknowledge the integrity of creation and creaturely causal capacities. The question is whether there ever exists one cause for complex forms of biological life, animal or human, and whether we will ever reach a point at which complex multilevel systems find a complete causal explanation. Rather, multilevel systems biology suggests that we need to think in terms of a *causal pluralism*, if we want to tackle multilevel complex systems. In terms of a theological response to multilevel systems I would rather propose a three-fold theological hypothesis of evolutionary systems. The first point has already been presented in my earlier response to strict neo-Darwinism, and the basic structure of the second point also follows the same trajectory as earlier presented, though with new material. The third point, however, is a theological view which I believe is resonant with multi-level systems biology alone, but also, of course, resonant if there is an interest in establishing a relationship between aspects of Christian theology and aspects of systems biology:

(1) *Metaphysical explanation*: God is the ultimate grounding source of all material reality and of all evolutionary possibilities. This is what Christian theology refers to as the most fundamental creation of mass-energy-information 'out of nothing' (ex *nihilo*).

(2) *Theological explanation from the principle of co-creativity*: Within the framework of the world of creation, God is everywhere at work as the enabling power that brings forth new evolutionary constellations but always within the evolutionary phase space. God is never a separate

'cause' but the enabling cause that works through underlying laws of nature, creaturely capacities, causal networks, and biological agencies. In this view, divine creativity always operates through the channels of evolutionary creativity, 'out of previously existing matter' (ex *vetere*) and 'out of the womb of evolving possibility' (ex *ovo*).

(3) *Theological explanation based on the outcomes of evolution:* The theological hypothesis here is that God facilitates and wants to create a world of intense life experience, facilitated by the natural processes indicated above. Accordingly, we see a biological world characterised by trends toward creativity, complexification, cooperation, sentience, aesthetic and moral values, and personhood, all in the understanding that it should become possible to live new forms of existence in trust (faith), forwards-orientation (hope), and other-directedness (love). Human forms of existence do not only grow out of the womb of evolutionary possibility (ex *ovo*) but also present intentional options for new ways of existence (de *novo*), facilitated by the mental and symbol realms of brains, minds, and cultural norms.

Contrary to the expectation of a worldview mapped upon a strict neo-Darwinian biology, we thus find in multilevel systems biology a sense of the wider spectrum of evolutionary outcomes, including cooperativity and intentional altruism. From such life experiences it is possible to recognise in the midst of the world of creation ways of life that are not dissimilar to what is religiously assumed to characterise divine nature. Life is about passionate love; but in passionate love there always looms the risk of pain and suffering. Self-sacrificial attitudes, wherever they occur, pay the costs of cooperation, and may even do so from an inner motivation.

Theologically speaking, creation is made for the sake of worldly fecundity and love, as the world was already created out of divine overflow of love (ex *amore*). But some ways of love are costly, and require a willingness to suffer for the sake of others. In this sense, not only externally adduced hardship is built into the story of evolution but also the intentional option of suffering for others. Accordingly, the story of evolution is not one of preordered harmony by design, but rather one of passionate love in all its meanings: struggling, giving, self-giving, giving in, and giving oneself up for others.

References

1. *Pew Research Center.* (2017) The changing global religious landscape. April 5 http://www.pewresearch.org/fact-tank/2017/04/05/christians-remain-worlds-largest-religious-group-but-they-are-declining-in-europe/ft_17–04–05_projectionsupdate_globalpop640px/ [31 December 2018].
2. Miller JD, Eugenie CS and Okamoto S. (2006) Public acceptance of evolution *Science* **313**: 765–766.
3. Haarsma D. (2017) New Gallup poll shows significant gains for the BioLogos view *Biologos.org*, May 24 [3 December 2018].
4. Depew DJ and Weber BH. (1995) *Darwinism Evolving. Systems Dynamics and the Genealogy of Natural Selection.* MIT Press, Cambridge, Mass.
5. Dawkins R. (2006) *The God Delusion.* Houghton Mifflin, Boston.
6. Gould SJ. (2002) *Rock of Ages: Science and Religion in the Fullness of Life.* Ballantine Books, New York.
7. Nowak M. (2014) God and Evolution In: Welker M (ed.) *The Science and Religion Dialogue. Past and Future*, pp. 47–52. Peter Lang, Frankfurt.
8. Holmes R. (2014) Care on Earth: Generating informed concern. In: Davies P and Gregersen NH (eds.) *Information and the Nature of Reality* (Canto Classics) pp. 262–312. Cambridge University Press, Cambridge.
9. Ruse M. (2001) *Can a Darwinian be a Christian? The Relationship between Science and Religion.* Cambridge University Press, Cambridge.
10. Wilson EO. (2007) *The Creation: An Appeal to Save Life on Earth.* W.W. Norton, New York.
11. Bråkenhielm CR. (2017) *The Study of Science and Religion. Sociological, Theological, and Philosophical Perspectives.* Pickwick Publications, Eugen, Oregon.
12. Gregersen NH. (2013) Evolutionary Theology. In: Runehov ALC and Oviedo L (eds.) *Encyclopedia of Sciences and Religions vol 2*, pp. 809–817. Springer, Dordrecht.
13. Guessoum N. (2011) *Islam's Quantum Question: Reconciling Muslim Tradition and Modern Science.* I.B. Tauris, London.

14. van Huyssteen W. (2006) *Alone in the World? Human Uniqueness in Science and Theology* (The Gifford Lectures 2004). Eerdmans Publishing, Grand Rapids.
15. Riedl R. (1976) *Strategie des Genesis*. Pieper, München.
16. Gregersen NH. (2001) The cross of Christ in an evolutionary world. *Dialog-J Theology* **40**(3): 192–207.
17. Southgate C. (2008) *The Groaning of Creation: God, Evolution, and the Problem of Evil*. Westminster John Knox, Louisville, KY.
18. Depew DJ. (2019) Organisms, development, and evolution: Invitation to a new understanding. *This volume*.
19. Crick FHC. (1970) The central dogma of molecular biology. *Nature* **227**: 561–563.
20. Mayr E. (1982) *The Growth of Biological Thought: Diversity, Evolution, and Inheritance*. The Belknap Press of Harvard University Press, Cambridge, Mass.
21. Gregersen NH. (1994) Theology in a neo-Darwinian world. *Studia Theologica* **48**: 125–149.
22. The Human Microbiome Project Consortium (2012) A framework for human microbiome research. *Nature* **486**: 215–221.
23. Monod F. (1972) *Chance and Necessity*. Collins, London.
24. Peacocke A. (2004) *Evolution: The Disguised Friend of Faith? Selected Essays*. Templeton Foundation Press, Philadelphia, PA.
25. McShea DW. (1998) Complexity and evolution: What everybody knows. In: Hull DL and Ruse M (eds.) *The Philosophy of Biology* (Oxford Readings in Philosophy), pp. 625–649. Oxford University Press, Oxford.
26. McShea DW and Brandon R. (2010) *Biology's First Law: The Tendency for Diversity and Complexity to Increase in Evolutionary Systems*. The University of Chicago Press, Chicago.
27. Wolpert DH. (2013) Information width: A way for the second law to increase complexity. In: Lineweaver CH, Davies P and Ruse M (eds.) *Complexity and the Arrow of Time*, pp. 246–275. Cambridge University Press, Cambridge.
28. Kauffman SA. (2013) Evolution beyond Newton, Darwin, and entailing law: The origin of complexity in the evolving biosphere. In: Lineweaver CH, Davies P and Ruse M (eds.) *Complexity and the Arrow of Time*, pp. 162–190. Cambridge University Press, Cambridge.
29. Jablonka E and Lamb MJ. (2005) *Evolution in Four Dimensions: Genetic, Epigenetic, Behavioural, and Symbolic Variation in the History of Life*. MIT Press, Cambridge, Mass.
30. Gatherer D. (2019) Modelling versus realisation: Rival philosophies of computational theory in systems biology. *This volume*.

31. Leyser O and Wiseman H. (2019) Integrative biology — parts, wholes, levels, and systems. *This volume.*

32. Wiseman H. (2017) Systems biology and predictive neuroscience. *Zygon: Journal of Religion and Science* **52**(2): 516–537.

33. Gregersen NH. (2017) The exploration of ecospace: Extending or supplementing the neo-Darwinian paradigm? *Zygon: Journal of Religion and Science* **52**(2): 561–586.

34. Laland KNF Odling-Smee J and Feldman MW. (2000) Cultural niche construction and human evolution. *Behav Brain Sci* **23**(2000): 131–175.

35. Nowak M with Higfield R. (2011) *Super Cooperators. Evolution, Altruism and Human Behavior, or Why We Need Each Other to Succeed.* Canongate, Edinburgh.

36. Sober E and Wilson DS. (1998) *Unto Others: The Evolution and Psychology of Unselfish Behavior.* Harvard University Press, Cambridge, Mass.

37. Coakley S and Nowak MA. (2013) Introduction: Why cooperation makes a difference. In: Nowak MA and Coakley S. (eds.) *Evolution, Games, and God: The Principle of Cooperation,* pp. 1–34. Harvard University Press, Cambridge, Mass.

38. Nowak MA. (2013) Five rules for the evolution of cooperation. In: Nowak MA and Coakley S (eds.) *Evolution, Games, and God: The Principle of Cooperation,* pp. 99–114. Harvard University Press, Cambridge, Mass.

39. Clayton P. (2013) Evolution, altruism, and God: Why the levels of emergent complexity matter. In: Nowak MA and Coakley S (eds.) *Evolution, Games, and God: The Principle of Cooperation,* pp. 243–361. Harvard University Press, Cambridge, Mass.

12 The Nature of Public Understandings of Biology: Between Comprehension and Dialogue

Steven Yearley*

What do we Mean by the Public Understanding of Biology?

Humans are chemical and physical beings, of course. Our cells pump protons at the same time as undertaking stupendous chemical syntheses. But there is understandably a greater fascination with our biological nature than with our chemical selves. This is, in part, due to the great life-science novelties of the last twenty years or so, when we have learned to talk about our genomes and, in many cases, signed up to companies that — for a fee — offer to inform us about our genomic and genetic characteristics. But it also relates to an older concern about exactly which of our attributes and dispositions should be understood in terms of our biological make-up. Although it is coming to be appreciated that we cannot really pose the question in such simple terms, it is still tempting to ask (for example) if gender characteristics

* Director IASH, Institute for Advanced Studies in the Humanities, University of Edinburgh, Hope Park Square, Edinburgh EH8 9NW, UK.

are primarily biological or cultural, and whether our individual psychological traits come from our biology or our upbringing. In this context, social scientists have come to reflect on the public's understanding of biology in the contemporary milieu.

At first sight, the public's understanding of biology (henceforth, for convenience, PUB) appears quite an obvious thing. It can be thought of as a measure of the extent to which ordinary members of the public know about, understand and accept the agreed interpretations of biological phenomena as they are developed by trained biologists. Of course, having that idea of what PUB is does not mean that one can measure or actually assess PUB in the UK or in the world as a whole. In order to get an empirical handle on PUB, one has to have a reasonable way of measuring it, one needs to have actually measured it, and one had better have some comparative yardstick to see how much biology is understood relative to say, astronomy or Olympic figure-skating.

Teachers, academics, research scientists, doctors and journalists have probably all along had some sort of feel for how biological issues or science in general are understood by some members of the public — the ones they meet or teach or talk to. But attempts to measure the state of public understanding first became well known in the mid-1980s and in the UK. This arose in the context of the Bodmer report (named after its chair, Sir Walter Bodmer, a celebrated geneticist) compiled for the Royal Society of London in 1985.[1,2] Bodmer's report on the Public Understanding of Science became key in establishing the 'Public Understanding of ...' framing as important to policy and politics. Essentially, what the Bodmer report did was to argue that the public's limited and weak understanding of science was of practical importance during a period of privatisation and the rise of government through market forces.

Under the free-market conditions of 1980s Britain, the prevailing policy assumption was that the market (people's revealed preferences) would indicate what was in demand — whether goods or services or advice. Producers should, accordingly, generate more of what was desired. At the same time, the government wanted to decrease its role in directing and funding scientific research since it believed that firms could be encouraged to underwrite the research they reckoned they needed; governments had no business second-guessing the needs of enterprises. In consequence, the scientific establishment suspected that it was receiving less support from the government than before, and that the free market in public preferences would mean that, for example, 'alternative therapies' might be treated on a par with scientifically validated medical research. These two trends might even be mutually reinforcing: if the public was not interested in official scientific opinion, that was one more reason for the government not to support science. And, if the government was not a strong, public supporter of science, this might lead ordinary citizens to think that they did not need to attend to scientists either.

Hence, the Royal Society's report argued that attempts to increase the public understanding of science (sometimes referred to as 'PUS') were practically important because, if people knew why scientific knowledge was robust and if people knew what scientists believed about specific issues, then the public would be more likely to take that advice seriously. The Bodmer Report urged that promoting the public's understanding of science was a duty of scientific researchers and should be adopted as a practical obligation by scientists funded out of the public purse as well as by other members of the scientific community.

In the late 1980s the social sciences in Britain were also feeling friendless, and the relevant research-funding agency (which had recently been renamed the Economic and Social Research Council — no longer a 'social *science*' council) saw an opportunity. It agreed to fund survey and other empirical work on the topic of the public's understanding of science to see whether the Royal Society's diagnosis was accurate. The specific novelty of the survey work undertaken in the wake of this proposal was that it included a 'quiz' element, actually testing the extent of the knowledge of a representative sample of the UK public.

In order to achieve this, there is one very important methodological step. To be able to correlate the results of the quiz scores with other variables that described the population (women versus men, older versus younger respondents and so on) one needs to generate variation among the quiz scores. That is, the quiz element had to include some questions which many people got right along with others that were answered correctly by relatively few people. If everyone got, say, 12 out of 20 answers correct then there would be no variation to correlate against the other variables. Accordingly, the quiz questions were extensively checked beforehand to be as sure as possible that there would be a good range of scores among the population tested. In that sense, the questions were never designed as a test of public knowledge — they were established in order to generate a range of scores. As the investigators note[3] (p. 12), the majority of the questions asked 'were selected to fall within the range of significant variation in public understanding of science'.

For practical reasons, the quiz could not be too long. But within the twenty-three questions originally asked in the main survey plus others in a component designed to test respondents' understanding of science's methods, there were in fact a number of biological and medical questions. Respondents were asked to mark as true or false or 'don't know' a set of statements which included ones about the number of legs that insects have, about plants as a source of oxygen, whether or not urine is made in the liver, whether the 'future children of a body builder will inherit the benefits of his training', and whether it is the father's 'gene' that determines if a baby will be a girl or a boy.

From the point of view of biologists, the results of these questions were pretty reassuring. Over 80% of respondents were right about the insects, almost 60% about the role of plants in generating atmospheric oxygen, over half about the liver, over half again about the X and Y chromosomes, and over three-quarters about the inheritance of acquired characteristics[3] (p. 14).

When the initial findings from the study were published and highlighted in an article in *Nature*[3,4] the standard socio-demographic results from the research were a little predictable. Gender and age-group differences, for example, were present, though not great, and the strongest correlation with quiz performance was length of exposure to education. This was reassuring for educators, but more or less what one would have predicted.

However, rather than the sociological findings garnering most attention in this flagship publication, the paper was framed around the idea that publics in the UK and USA are weak in their average understanding of science: 'how much science does the general public understand? The answer is ... "not much"'[3] (p. 11). To underscore this conclusion the focus was turned onto one particular question: the one that asked whether the Earth goes round the Sun or vice versa and how long it takes to do so. Only 34% of UK participants got this fully correct (i.e. the Earth travels around the Sun and it takes a year to do so[a]), leading the authors to note that educators have not passed on enough information to allow the public to have 'caught up with Nicholas Copernicus and Galileo Galilei'. Indeed: 'If modern science is our greatest cultural achievement, then it is one of which most members or our culture are very largely ignorant'[3] (p. 13).

The paper in *Nature* presented the findings as reflecting very poorly on what the British public knew and suggested that the scores on what they now termed the 'Oxford Scientific Knowledge Scale' were disappointingly low. The questions had also been adapted for a USA/UK comparison where the score of the US respondents was slightly better on average. The scale was subsequently used in a European comparison where the UK performed relatively strongly[3,5] (p.12; p.169).

From this framing of the results came the idea of a public 'deficit' in scientific knowledge which has now become part of accepted parlance about this issue. It was all too easy to represent the findings — as the *Nature* article tended to do — as showing an alarming lack of scientific understanding amongst the UK population. Given the pivotal historical role

---

[a]On a yet more sophisticated view of the matter, the Earth does not so much go around the Sun (as if the Sun were stationary) as move with it. Still, it seems unlikely that anyone got the question 'wrong' because of taking this more advanced view.

played by arguments over geocentric versus heliocentric models of the planetary system there are, perhaps, arguments for taking the Copernican question as the core one. Yet — as noted above — the population did rather better on biological questions which could have been used to counter the main storyline. However, the focus on deficits quickly developed as a way of interpreting the entire issue. If one sees the survey as demonstrating widespread public ignorance, then it is a short step to assuming that, if only people knew more about science, they would take scientists' authority more seriously.

Reasoning Using Deficits

The pattern of using evidence for a deficit in public comprehension as a way of gauging the cultural standing of one's own enterprise has continued to the present. In 2015 the Royal Society of Chemistry commissioned a UK study of the public understanding of chemistry. Chemists, in Britain at least, face a tricky semantic problem of being confused with pharmacists since pharmaceutical outlets are still routinely known by some native British English users as 'chemists', as are the professionals who work there. The Royal Society of Chemistry was also interested in the sense in which members of the public appeared to be concerned about things being made of 'chemicals' when — on a scientist's view of the world — there are no everyday objects not made of chemicals. Happily, 60% of respondents agreed that 'everything is made of chemicals' though the report also chose to highlight the finding that people 'don't have an emotional connection with chemistry'.

There is a growing interest in the public understanding of history and recently the UK's Royal Society of Literature commissioned an opinion poll on Literature in Britain Today.[6] This focused most on people's interest in literature and on the benefits associated with reading but did have a quiz element with 20% of respondents acknowledging that they cannot name any author of literature — not even JK Rowling[6] (p. 23).

Despite warnings about not deploying the language of deficit too readily, it is frequently mobilised by prominent scientists. A particularly clear example was offered about a decade ago (but long after Bodmer) by Steven Pinker in a *New York Times* review of a book by science writer Natalie Angier (*The Canon: A Whirligig Tour of the Beautiful Basics of Science*).[7] Apropos the need for this kind of book, Pinker wrote:

> A baby sucks on a pencil and her panicky mother fears the child will get lead poisoning. A politician argues that hydrogen can replace fossil fuels as our nation's energy source. A consumer tells a reporter that she refuses to eat tomatoes that have genes in them. And a newsmagazine condemns the prospects of cloning because it could mass-produce an army of zombies.
>
> These are just a few examples of scientific illiteracy — inane misconceptions that could have been avoided with a smidgen of freshman science. (For those afraid to ask: pencil 'lead' is carbon; hydrogen fuel takes more energy to produce than it releases; all living things contain genes; a clone is just a twin.) Though we live in an era of stunning scientific understanding, all too often the average educated person will have none of it. People who would sneer at the vulgarian who has never read Virginia Woolf will insouciantly boast of their ignorance of basic physics. Most of our intellectual magazines discuss science only when it bears on their political concerns or when they can portray science as just another political arena.[8]

The claim here could hardly be clearer. Though we live in an era where 'stunning scientific understanding' is available, many otherwise-educated people are happy to be ignorant of science. This leads them to be fearful about the wrong things and to make unwise policy or consumer choices. Moreover, an ignorance of science is not something about which people are encouraged to feel shame; on the contrary it is culturally tolerated and endorsed. These points can then be developed into calls for changes to the media, for alterations in the attitudes of political leaders and policy makers, or for reforms in science funding structures to offer incentives for scientists to engage in more public outreach.

This is a prime example of using the public 'deficit' as an explanatory resource or trope. But the strategy is not that easy to deploy successfully. For one thing, despite Pinker's undoubted status as a leading scientific thinker, even his training and immersion in the culture of science does not protect his writing from deficiencies. The production of hydrogen fuel clearly costs energy, but if, say, the electricity itself is from renewable resources, then it is quite thinkable for hydrogen to power transport. It is also hard to sustain the idea that a clone is ordinarily 'just a twin' since notable clone mammals have been cloned from adult cells, making them rather different from ordinary twins. It is hard to be sure where the deficits start and stop.

Indeed, right from the start, reasons had been offered for not charging down this particular 'deficit' path. First of all, as noted above, there were other questions in surveys where public respondents did better, where there was less obvious evidence of ignorance. It was the analysts' decision to elevate particular questions into the canonical ones. Moreover, the quiz format does not really discriminate between cases where respondents don't know or are confused about scientists' claims with occasions when they know what scientists think but just happen to disagree. For example, some of the US respondents answered that dinosaurs and early people occupied the Earth at the same time. They may just have got this mixed up or they may actively believe this on religious grounds, viewing this as one of the rare occasions when scientists have got things wrong. The same applies to questions about the difference between synthesised and 'natural' vitamins; members of the public may know what scientists are likely to think about this but just arrive at a different conclusion themselves.

The authors of the *Nature* PUS article themselves quickly stepped back from strong versions of the deficit claims. Using the same results, they subsequently noted that:

> the differences between the well informed and the less well informed with respect to the degree of discrimination between particular areas of scientific research — and especially the tendency for the well informed to express less support for morally

> contentious research — suggests that it would be unwise for
> scientists and science policy-makers to presume that a better
> informed public is automatically a public that is more supportive
> of any and all forms of scientific research.[9] (p. 70)

That is to say, on close inspection, greater scientific knowledgeability among respondents did not seem to correlate with higher support for scientific authority. People who displayed the most scientific literacy seemed to be more selective in their support for scientific authority. This general relationship also seemed to be repeated at a national level, with citizens from the most technically and scientifically advanced nations being less uniformly supportive of scientific authority than respondents from somewhat less technically and scientifically advanced countries[5] (p. 170 and 175). Greater facility with scientific ideas may be associated with critical attitudes towards scientific developments.

Although these difficulties with 'deficit thinking' were becoming more and more apparent, it seemed that it was hard to shake off the trope. Researchers and consultancies were still being hired to assess the public understanding of this and that. The group that undertook the study for the Royal Society of Chemistry sought to address this issue head-on, asserting that they were:

> moving away from what is termed the 'deficit model' of public
> attitudes towards science, to a new way of approaching interac-
> tion between the sciences and the public. The 'deficit model'
> is the idea that public concerns and scepticism about science
> and scientific developments can only be the result of a lack of
> understanding, and can thus be countered with [sic] providing
> people with rational evidence and information. In other words,
> it is to take the position that if only people understood the sci-
> ence, they wouldn't be worried about it. Current approaches
> instead characterise the relationship much more in terms of
> the broader social implications of science, and do not attrib-
> ute public concerns to cognitive deficit. Rather these concerns
> are understood as the product of genuine moral, social and

political deliberations, questions of what is ethically acceptable, about who is affected by science and in what ways, who
benefits and who makes money, and at what expense. Modern
science communication accepts that such concerns cannot be
dismissed with scientific facts.[10] (p. 6)

Whether such 'modern science communication' approaches to surveys of public knowledge were going to be sufficient or not never came
to be really tested because other responses to the deficit issue came to
outflank this strategy.

A Dialectic of Ignorance and Knowledge

In order to argue against the assumption of widespread public ignorance
and the presumed connection between such ignorance and distrust or
rejection of scientific authority, one attractive and logically robust strategy
was to identify and document knowledgeable publics. Qualitative case
studies (including those supported by the Economic and Social Research
Council in the same programme as the survey highlighted above) have
been critical here. Such studies indicated that, in instances where scientific
knowledge was salient to communities or families, people might already
possess knowledge that was empirically robust or could acquire it relatively
rapidly. These citizens might not score well on a science quiz but that did
not mean they were not well informed about aspects of science related to
their lives or interests.

For example, persistent public worries about the incompleteness of
official measures of nuclear safety around the Sellafield plant in north-west
England, coupled with the work of concerned scientists, led to closer attention to the subtle biological pathways through which radioactive contamination can be concentrated[11] (pp. 290–295). Public concerns were not so
much evidence of a deficit; the deficiency was rather more on the side of
regulatory scientists who had not investigated all the possible pathways.
In a slightly later study of air quality monitoring in the northern English city

of Sheffield, local people were interviewed in focus groups representing different stakes in air quality and pollution: traffic campaigners, residents' associations, cycling groups and so on. Engaging with outputs from the official (council-run) model, cyclists and traffic campaigners maintained that street-level pollution was far more variable than the model indicated. They urged that their experience indicated that it was concentrated around bus routes with buses powered by specific (ageing) engines. In a way unknown to the model, the air quality tended to be (and was subsequently verified as being) worst around those buses' routes and particularly in the areas where they idled, waiting to get back onto the timetabled routine[12] (pp. 115–116). Similarly, patient groups who have insiders' knowledge of a disease or disorder have contributed to understanding how to manage their condition, particularly on how to regulate it not in standardised ways but in the light of the varied and unpredictable demands of everyday family life.[13] In these ways, lay publics can be seen to have been active participants in the generation of new knowledge and the overthrow of old scientific beliefs. The relationship between scientific expertise and the public is far more complex than is typically recognised in the calls for 'public understanding' which emanate from the scientific establishment.

These studies also tended to place new emphasis on the role of trust in the public assessment of scientists' claims. Given that judgements about trustworthiness are central to the practice of science (as Shapin[14] has skilfully emphasised), there is little hope that they can be eliminated from the processes by which the public comes to acquire and assess scientific knowledge. Trust is an indispensable component in the creation and pass-ing on of scientific knowledge; it is not a feature restricted to lay audiences for science that can be technically manipulated to promote 'high trust' conditions. Furthermore, trust and credibility are not fixed dispositions; they are the outcome of interactions and negotiations between citizens and the scientific experts they encounter.

Overall, these insights from case studies in the public understanding of science revealed two main sorts of conclusions. First, they indicated that there is no one formula for transmitting scientific knowledge. The credibility of experts is in a sense always being negotiated and evaluated. A means of deploying expertise in one social context may not work in another. This is because public trust in expertise is a (perhaps the) central issue in PUS, and trust cannot readily be routinised. Secondly — and more radically — it is not accurate or appropriate to regard the public understanding of science as a one-way traffic.

To show that people possessed knowledge which could 'stand up to' and even contribute to recognised experts' understanding was sufficient to argue against the 'deficit' view of the public as simply ignorant. But the rejection of ignorance led to a corresponding query about what public knowledge amounts to. In a dialectical manner, to deny that someone is deficient in knowledge or to deny their ignorance is — at the same time — to imply that they know something. This raised a problem of what to make of this knowledge; it also opened a possibility that this public knowledge could be harnessed in some useful way.

Clearly, one does not want to move from this new conception of the public as 'not ignorant' to a romantic notion that the public is widely knowledgeable about recondite aspects of science, nor — for that matter — about economics or fine art. But a reluctance to treat the public as ignorant could readily turn out to be equivalent of regarding publics as possessed of relevant common sense or as in some way generically wise.

One outcome of this dialectical kind of argumentation was new attempts to codify the kinds of knowledge that different types of actor could bring to a policy issue or practical problem. Within the sociology of science literature, one version of this enterprise elaborated by Collins and Evans[15] (pp. 13–35) has become relatively well known. They have sought to outline a so-called 'periodic table' of expertises, distinguishing (for example) between someone (maybe a science journalist or policy maker) who can talk fluently about a technical area and someone who has the expertise to make new, recognised contributions to the research front. They were keen to counter the flattening-out apparently introduced by discovering the knowledgeability of everyday actors and aimed to clarify the kinds of knowledge and insight that different classes of actors could offer.

A second approach aimed less to codify the kinds of knowledge available and more to devise a way to take advantage of the benefits afforded by the acknowledgement of the public's insights.[16] Funtowicz and Ravetz proposed that, rather than taking the public as co-creators of knowledge, it would be preferable to devise forms of 'extended peer review'. This would be particularly the case in relation to topics, such as climate change, where there are very high stakes and significant uncertainties; in such instances, disciplinary experts are necessarily limited in their insights and there is the need for the exercise of judgement. The extended peer review would engage members of the public as reviewers of scientific knowledge claims, essentially involving citizens in the quality assurance process.

Significant though these attempts to develop an interpretative framework for public knowledge have been, more influential were practical steps to take account of public knowledge that were hastened by the UK House of Lords report on Science and Society (2000).[17] This report legitimated at the highest political level the idea that it would be wise and practicable to involve citizens in the assessment of scientific and technical matters. And in the 2000s several high-profile biological-science-related issues became the focus for new forms of public consultation and evaluation, including genetically modified organisms (primarily crops and foodstuffs), synthetic biology and stem cell biology. PUB was suddenly back.

What Role for Public Knowledge in PUB?

In the new context of the early 21st century, there was a willingness to engage the public in the assessment of science for policy purposes. There was, in principle at least, a drive to distance these exercises from the old deficit model. And there were many new biological issues to be addressed.

However, the new biological developments were, by definition, things which were unfamiliar to most people. Indeed, in some cases they were related to highly innovative and complicated ideas far removed from people's everyday experiences. With new possibilities such as in stem cell research and synthetic biology, gene drives and CRISPR, there were of course 'deficits' in the sense that most people would not know what pluripotent stem cells are before such cells were in the news or described in policy arguments over the rights and wrongs of developing early-stage embryos for medical research.[18] In practice, this meant that survey and questionnaire work on PUB issues in this period tended to return to asking about people's attitudes.[19,20] Gaskell et al.'s[21] European study did contain a small quiz element but again the main focus of the study was on changes in attitudes towards new biotechnologies.

From this difficulty arose a second problematic issue. It was unclear precisely what it was that citizens were expected to bring to this encounter with new biological sciences and by virtue of which qualifications ordinary people were being consulted. In the 'Synthetic Biology Dialogue'[22] — an expensive and large-scale British exercise supported by the two main UK Research Councils (the BBSRC and EPSRC) — groups of citizens were recruited in regions around the UK and brought together to discuss options for the technology, its regulation and development. In such exercises, painstaking and well thought through as they were, it is clear that citizens are not being recruited because they are deemed to be knowledgeable about the scientific issues; on the contrary, the fundamental biological issues were outlined during the early parts of the first meeting. Rather, they are to be involved because of their assumed insights about the ethics of taking certain kinds of technological risks or about people's willingness to accept certain kinds of innovative products. In practice, such processes can tend to reintroduce some of the previous asymmetry between scientists and the public, except that there is now acceptance that public consultation and engagement is a necessary step in policy making.

This leads to one final, related question: what power is granted to the dialogue exercises? In other words, people may be involved in discussions over synthetic biology or stem cell science futures, but what kind of presumption is there about the ways in which their views will come in the end to be reflected in policy? Horlick-Jones *et al.*[23] carried out an external assessment of the British 'GM Nation?' exercise which ran alongside the public consultations over genetic modification that had taken place in 2003. These authors pointed out that it was difficult to get an 'authentic' public consultation to guide policy since there were questions over the representativeness of the 'public' that got to participate; furthermore, the process and the kinds of conclusions that could be offered were shaped by its instigators not by the participants. On a smaller scale, Matthew Harvey[24] undertook a participant observation study of a subset of the GM Nation? groups. He not only confirmed that participants seemed unclear about their relation to policy-making over GMOs, he also focused on participants' experience of the exercise and on the kinds of topics participants opted to talk about. He noted that the public participants' discussion often turned on scientific questions rather than on public-interest themes. Given a more or less free rein, he observed, public participants chose to argue about such issues as safe planting distances, the impact of GM crops on beneficial insects and so on, rather than on the ethical and socio-economic issues on which they were assumed to be informed and by virtue of which their participation was officially legitimated. On Harvey's interpretation, many participants felt caught in endless and frustrating debates which they were typically unable to resolve in the context of the meetings.

There is a forceful irony here. In the last two decades PUB has moved from a topic of merely academic enquiry to a prominent feature of key policy consultations around scientific and technological matters. However, these consultations have typically been designed around the idea that the public brings ethical or consumer or even economic insights to the questions at hand. Roughly speaking, biologists tell us what it is possible to do and the public is consulted over whether or not it should be done (on grounds of moral acceptability or equity and so on). For this, the public's substantive biological knowledge is unimportant. Out of a focus on the public understanding of science has grown a reasonably well institutionalised practice of public consultation for which the public's understanding of biology is pretty much beside the point.

Conclusions

This chapter has investigated the evidence on the way that biological knowledge is understood and interpreted in the UK and other contemporary Western societies; it has also reviewed and discussed new roles that have been devised for public engagement with contemporary biology, for example in synthetic biology and in stem cell research.

The early work in PUS appeared to demonstrate major public knowledge deficits, even if those deficits did not seem to have much explanatory power in terms of accounting for differences in public acceptance of scientific authority. However, on these 'classic' deficit studies, the public performance on the biological questions was comparatively good. Had biological topics been chosen as the archetypal scientific questions, the headline news would likely have been different: British citizens would have appeared reasonably knowledgeable and scientifically literate.

Attempts to counter the dominant framing of the public's deficient understanding of science led to initiatives designed to register the ways in which members of the public were in fact knowledgeable. This, in turn, led to novel proposals for mobilising public knowledge for research and policy purposes.

In the 21st century, these strategies have come to greater prominence in the biological area than in most other scientific domains. They were adopted in several leading areas of innovative biological science and biotechnology. In several of these areas — including stem cell biology and synthetic biology — the innovations were profound and not closely related to widespread, pre-existing public knowledge. In these cases, members of the public (and policy makers and even scientists from other disciplines) were characterised by knowledge deficits. People did not know what pluripotent stem cells were before these cells featured in the news. Under these circumstances, the presumed basis for public participation shifted subtly. Members of the public were engaged not because of the biological or biologically related insights they might have, but on account of the role they could play as representatives of the public's ethical sensibilities and citizens' willingness to embrace new products or processes. Several studies and reviews suggested that this was not a role with which they felt routinely comfortable. These ventures — though commonly claiming to distance themselves from previous 'deficit approaches' — made little use of public knowledge. Indeed, this tended to happen rather less than in some 'Citizen Science' projects which did find novel ways to work with knowledgeable publics (see, for example, Ellis and Waterton).[25] The enduring challenge for PUB is to design methods of participation in policy debates which benefit from the kinds of knowledge people can meaningfully bring.

References

1. Bodmer WF. (1985) *The Public Understanding of Science. Report of a Royal Society ad hoc Group Endorsed by the Council of the Royal Society.* Royal Society, London.
2. Bodmer WF. (1987) The public understanding of science. *Science and Public Affairs* **2**: 69–90.
3. Durant JR, Evans GA and Thomas GP. (1989) The public understanding of science. *Nature* **340**: 11–14.
4. Evans GA and Durant J. (1989) Understanding of science in Britain and the USA. In: Jowell R, Witherspoon S and Brook L (eds.) *British Social Attitudes, 6th, Report*, pp. 105–120. Gower, Aldershot.
5. Bauer M, Durant J and Evans G. (1994) European Public Perceptions of Science. *Int J Public Opin R* **6**(2): 163–186.
6. *Royal Society of Literature* (2017) Literature in Britain today. https://rsliterature.org/wp-content/uploads/2017/02/RSL-Literature-in-Britain-Today_01.03.17.pdf.
7. Angier N. (2007) *The Canon: A Whirligig Tour of the Beautiful Basics of Science.* Houghton Mifflin, Boston, MA.
8. Pinker S. (2007) Review of Angier *New York Times*. https://www.nytimes.com/2007/05/27/books/review/Pinker-t.html [27 May 2007].
9. Evans GA and Durant J. (1995) The relationship between knowledge and attitudes in the public understanding of science in Britain. *Public Understanding of Science* **4**(1): 57–74.
10. Royal Society of Chemistry (2015) Public Attitudes to Chemistry London: Royal Society of Chemistry.

11. Wynne B. (1992) Misunderstood misunderstanding: Social identities and public uptake of science. *Public Underst Sci* **1**: 281–304.

12. Yearley S. (2000) Making systematic sense of public discontents with expert knowledge: Two analytical approaches and a case study. *Public Underst Sci* **9**: 105–122.

13. Lambert H and Rose H. (1996) Disembodied knowledge? Making sense of medical science. In: Irwin A and Wynne B (eds.) *Misunderstanding Science? The Public Reconstruction of Science and Technology*, pp. 65–83. Cambridge University Press, Cambridge.

14. Shapin S. (1994) *A Social History of Truth: Civility and Science in Seventeenth-Century England*. University of Chicago Press, Chicago.

15. Collins H and Evans R. (2007) *Rethinking Expertise*. University of Chicago Press, Chicago.

16. Funtowicz SO and Ravetz JR. (1991) A new scientific methodology for global environmental issues. In: Costanza R (ed.) *Ecological Economics*, pp. 137–152. Columbia University Press, New York.

17. House of Lords Select Committee on Science and Technology (2000) *Third Report on Science and Society*. HMSO, London. https://publications.parliament.uk/pa/ld199900/ldselect/ldsctech/38/3802.htm.

18. Parry S. (2009) Stem cell scientists' discursive strategies for cognitive authority. *Sci Cult* **18**(1): 89–114.

19. Shepherd RJ, Barnett H, Cooper A, *et al.* (2007) Towards an understanding of British public attitudes concerning human cloning. *Soc Sci Med* **65**: 377–392.

20. Sturgis P, Cooper H, Fife-Schaw C and Shepherd R. (2004) Genomic society: Emerging public opinion. In: Park A, Curtice J, Thomson K, *et al.* (eds.) *British Social Attitudes: The 21st Report*, pp. 119–145. Sage, London.

21. Gaskell G, Allum N and Stares S. (2003) *Europeans and Biotechnology in 2002, Eurobarometer 58.0, A report to the EC Directorate General for Research from the project 'Life Sciences in European Society'*. European Commission, Brussels.

22. TNS-BMRB (2010) *Synthetic Biology Dialogue*. BBSRC, Swindon.

23. Horlick-Jones T, Walls J, Rowe G, *et al.* (2004) A deliberative future? An independent evaluation of the GM Nation? Public Debate about the possible commercialisation of transgenic crops in the UK, 2003. (Understanding Risk Working Paper 04–02). Centre for Environmental Risk, Norwich.

24. Harvey M. (2009) Drama, talk, and emotion: Omitted aspects of public participation *ST&HV* **34**(2): 139–161.

25. Ellis R and Waterton C. (2004) Environmental citizenship in the making: The participation of volunteer naturalists in UK biological recording and biodiversity policy. *Sci Publ Policy* **31**(2): 95–101.

13 How Children Understand Biology

Michael J. Reiss*

Introduction

As any parent or other observer of young children knows, in the first few years of our post-birth lives we go through an extraordinary transformation. Within the first few months, most babies begin to smile, to track objects with their eyes, to bring their hands to their mouths and to reach for dangling objects. Within the next three months, they typically mange to roll over when on their backs, babble, laugh, manipulate objects with their hands and sit up with support. By the time they reach their first birthday, many babies are now toddlers, able to move around on their feet while holding onto things, respond to familiar words, say a few words, point at objects in order to get attention and even take a few independent steps.

The focus of this chapter is on how children come to understand aspects of the natural world, specifically biology during their pre-school and school years. In a book on the public understanding of biology such a chapter is appropriate in its own right — children (young people) are themselves people. In addition, today's young people are tomorrow's adults.

*Professor of Science Education, Institute of Education, University College London, 20 Bedford Way, London WC1H 0AL, UK.

Furthermore, as a result of studies in psychology and biology education, we are beginning to build up a good understanding of how children understand biology and, perhaps more interestingly, why.

A key argument of this chapter is that young children have evolved to learn in ways that relate closely to their understanding of biology. They naturally categorise objects, including animals and plants, and have implicit models of causation that differ for animate and inanimate objects. This means that by the time they start their formal school education they are most definitely not *tabulae rasae*. Rather, they arrive at school with a number of beliefs, for instance that objects need to be supported to avoid falling downwards (so, a box falls if the table on which it is standing is withdrawn), that changes to inanimate objects are generally the result of causes whereas animals are agents who can move on their own (so, it is not surprising when a dog moves without anything coming into contact with it but it is when a meaningless shape does) and that artefacts do not grow but organisms do (so, small kettles do not increase in size over time whereas rabbits do).[1]

Children also arrive at school with biological notions that they acquire from their family and others; some of these will be scientifically accurate, others not so. Educators have adopted one of two approaches when children's views about scientific phenomena in general, or biological ones in particular, differ from the accepted views of scientists. Either they have labelled such views as 'misconceptions', with an implication that such views need changing, or they have labelled them as 'alternative conceptions', with an implication that they may contain a certain validity. There are advantages and disadvantages to each approach. Taken to extremes, the 'misconceptions' approach can give the impression that school biology is all about students simply learning what is correct and unlearning what is incorrect. This neither sounds like a very good way of enthusing students about their learning nor represents a robust understanding of the nature of scientific knowledge and how it changes over time. The 'alternative conceptions' approach, though, can give the mistaken impression that 'anything goes'. This isn't true in any intellectual discipline and certainly not in biology.

In the below I look at understandings and misunderstandings (perhaps a less value-laden term than misconceptions) in a number of areas of biology. Particular attention is paid to children's understanding of evolution and the reasons why such views may be resistant to change. At the same time, effective strategies for enabling children to learn the scientifically accepted account of evolution are discussed.

An Evolutionary Perspective on the Developing Child's Understanding of the Natural World

It might be thought that our perceptual apparatuses (sight, hearing, smell and so on) and our brain would have evolved to give us as accurate a view of the world as possible. Of course, we know, to take just the sense of hearing, that there are certain frequencies of sound that we cannot hear (for example, the high-pitched noises many bats make when using echolocation to obtain their food). Equally, we know that other species may have senses that are more acute than ours, for example the smell of a dog or the vision of an eagle.

But our senses are much less able to provide us with the truth of the external environment than is indicated by these relatively minor considerations.[1] For example, we hear silences between the words in a conversation whereas the information gathered by a microphone 'listening' to the same conversation shows that these silences are, objectively, full of sounds. The gaps we hear between words are very largely gaps inserted by our minds, in a process termed 'speech segmentation', to render the stream of auditory information more intelligible. Speech segmentation is an important ability that we acquire as we develop.

Then there is the fact that our visual system is far more sensitive, to give just one example, to movement than to non-movement. The evolutionary advantages of this are clear: a mother needs to be sensitive to changes in the behaviour of her child(ren) and changes are often indicated by movement; hunters need to be alert to the movements of possible prey; and so on. Indeed, we share this hypersensitivity to movement with many other species — the classic work was done on frogs back in the early 1950s.[2] This is an illustration of the fact that it is not only humans that sense only part of the world and emphasise other parts of it, so that we really do construct the world around us; this is a characteristic we share, to varying degrees, with all of life including the first unicellular organism that evolved the ability to detect the presence of certain chemicals in its surroundings, while 'ignoring' others, and so was able to spend more time in places where there was more food.

The simplest way of summarising all this is to say that the external world is a complex world and even the human mind, in all its richness and complexity, simply has neither the capacity nor the need to obtain a full, detailed map of it. Rather, we have evolved to get through life and leave offspring that have an above average chance of doing the same. The evolutionary perspective is not that we have evolved to discern truths but to succeed in surviving and reproducing.

What Sort of Mind has Evolution Given Us?

The rapidly advancing field of cognitive development is providing an ever-deepening understanding of the infant mind and of how it develops. There is much in here that is illumined by an evolutionary perspective. For example, we have a tendency to see animate shapes in non-material objects (e.g. clouds, hedges, Rorschach test cards) and to attribute intentionality to certain movements. The likely reason for such tendencies is that the evolutionary costs of failing to discern animate objects or intentionality when these don't, in actuality, exist are smaller than the evolutionary costs of failing to discern them when they do. If you think you hear a dangerous creature under your bed, the worst that usually happens is that it takes you longer to get to sleep. But such caution might save your life (to be more precise, might have done — rural Cambridgeshire, in my case, has few poisonous snakes or other creatures that lurk where I sleep).

Developments within cognitive psychology and the younger discipline of cognitive neuroscience over the last twenty years or so are providing us with a better understanding of what we might term the 'rules about the natural (including human) world' with which we are born or develop within the first few years of life. Such rules, for example about causality and that solid objects don't pass through one another or disappear and then reappear, can perhaps best be viewed as shortcuts; they enable all of us to interpret rapidly, from an early age and in a way that is useful to us, much of the mass of information our brains receive about the world about us — always remembering that from an evolutionary perspective utility is understood not by reference to ultimate truths but to the pragmatics of survival and reproduction.

A thorough overview of these 'rules' is provided by Usha Goswami.[3] Much of the evidence for these relies on the useful observation that babies and infants spend longer looking at the unexpected. (One can note, parenthetically that this itself is of evolutionary significance. By and large, it's the unexpected and the unfamiliar that deserve special attention.) So, for example, three and a half month old babies spend substantially less time staring at an experimental set up that shows a moving short carrot that disappears behind a short wall and then reappears than they spend staring at an experimental set up that shows a moving tall carrot disappearing behind a short wall and then reappearing. In everyday language they 'know' (and it is not my intention here to get into a discussion about whether such 'knowledge' is truly cognitive or simply the result of sensory perception) that part of the tall carrot should have been visible above the short wall.

Research on Children's Understanding of Biology

Policy documents in various countries all over the world often suggest that scientific literacy should be a major aim of education at schools. This is especially important for biology, as there are many socio-scientific questions that students will encounter during their lives, and which may demand relevant decisions, that require a good understanding of biology if they are to be answered. Such questions may be related to environmental issues, genetic testing, infectious diseases and more. Biological research in the 21st century has an impact on various aspects of human life, with important implications for how we understand health, disease and identity. Therefore, literacy in biology is a core component of scientific literacy. Future citizens who are literate about science should definitely have an appropriate understanding of biology for medical, environmental and other reasons.

However, research in teaching and learning of biology, as well as in its public understanding, reveals serious shortcomings.[4] It seems that — generally speaking — it is possible for many students to leave school holding inaccurate conceptions and having substantial misunderstandings of many biological phenomena. There are various reasons for this. For one thing, biological research advances at a fast pace and it is often difficult for biology teachers to keep up with it; then there is the fact, discussed above, that our minds have not evolved to have a good scientific understanding of the world but to survive and reproduce in it. Much of science is counter-intuitive and changing people's minds is not always straightforward.

In the rest of the chapter I spend more time on children's understandings of evolution than on any other biological topic but, to illustrate that misunderstandings in biology are not restricted to this issue, it is worth looking at some other topics too and discussing the reasons for misunderstandings.

Microbiology

Carson, Dawson and Venville[5] discuss children's understandings about microbiology. Microbiology is concerned with organisms (microbes), such as bacteria and viruses, that are too small to be seen with the naked eye. Perhaps the most important misunderstanding is the widespread presumption that all microbes are harmful.[6] This makes sense from an evolutionary perspective; as argued above, it generally benefits one to pay more attention to things that could be dangerous than benign. So, it is perhaps unsurprising that children equate microbes with 'germs',[7] seeing microbes as a source of infection and illness.

Any evolutionary tendency to see microbes as harmful is no doubt reinforced by the well-meaning (and entirely appropriate) advice we receive when young that we should wash our hands after going to the toilet, after coming indoors and before eating. Of course, we now know that there are dangers in taking such advice to extremes. A small proportion of people may develop compulsive hand washing, one of the most common manifestations of obsessive-compulsive disorder (and one for which there can be other triggers), and a greater proportion of people are more likely to develop various autoimmune diseases and allergies as a result of excessive cleanliness — the so-called 'hygiene hypothesis'.[8]

More mundanely, microbes are of value to us through such traditional food-making activities as bread-making and brewing, in the recycling of matter from organisms that have died and as a crucial, though still little understood, component of our guts. We would want a good school education to result in young people knowing something of the benefits that microbes bring to us as well as their harms. Of course, all this is a very anthropocentric way of regarding microbes; microbes are worth considering in their own right. This leads me to a consideration of ecology and what we might want a good school ecology education to achieve.

Ecology

Ecology, from the Greek *oikos*, home, is the study of the distribution and abundance of organisms including ourselves. It encompasses many subdisciplines as defined by the species that are studied (plant ecologists, animal ecologists, etc.), the communities that are studied (grassland ecologists, marine ecologists, etc.) or other considerations (molecular ecologists, urban ecologists, etc.).

In most countries, ecology has lagged behind other components of the biology curriculum in schools. When I started to teach biology in schools in 1983, there was still debate about the structure of the unit membrane (the layer that bounds all cells and many subcellular components — the debate was about whether the proteins formed a continuous layer on each side of the lipid bilayer or existed as occasional molecules that straddled the membrane, the fluid mosaic model) and whether mitochondria and chloroplasts had once been independent organisms. Nowadays, we teach in some depth about the ways in which proteins, in particular, are able to control the movement of molecules across such membranes and about the genetic evidence for endosymbiosis. Advances in molecular biology have been even greater. However, ecology not only seems not to have advanced, if anything it has receded. There are many reasons for this, some of which are self-perpetuating. Biology degrees are much less general than they used to be; many good biology graduates study no ecology while at university and are almost completely unfamiliar with common organisms in the environment — a high proportion of students training to be biology teachers are unable to recognise a hornbeam (*Carpinus*) or a lime tree (*Tilia*), a nuthatch (*Sitta europaea*) or a treecreeper (*Certhia familiaris*).

All this is particularly unfortunate since, as is widely acknowledged, ecology is likely to be a crucial area of biology in the lives of today's school students. We are living through such an unprecedented destruction of the world's biota that there is growing pressure for our age (epoch) to be named the Anthropocene; the evidence for anthropogenic climate change is increasing rapidly but the consequences (including the possibility of it being reversed) are unclear; it is not yet known whether biotechnology will help solve many environmental problems or only add to them. The concept of ecological literacy refers to the importance of educating citizens so that they not only know quite a bit about ecology but are able to process that knowledge and apply it for informed decision making and environmentally-related actions.[9] We need teachers and students with good understandings of ecology. We also need citizens who feel positively about the natural environment, want to preserve biodiversity and are prepared to act accordingly.

Students have difficulties in understanding the relationships organisms have with other species and with the abiotic elements of their environment. Indeed, student understandings of interactions in ecosystems do not improve much as they progress through their schooling.[10] In addition, it has been repeatedly reported that students tend to think in terms of isolated food chains rather than interconnected food webs.[11] Furthermore, when considering the effects of changes in population sizes, students tend to be able to more easily comprehend the effects of changes up the trophic levels (e.g. primary consumer (herbivore) → secondary consumer (carnivore)) than down.[12]

Students are also generally unaware of how feedback mechanisms operate in ecosystems. For example, in a hypothetical predator–prey system of wolves and deer, students usually believe that the wolves will first consume all the deer and then themselves die, since they will have exhausted their feeding sources.[11] Other students, though, have a great belief in the 'balance of nature', holding that no matter what the kind or the magnitude of the disturbance, the ecosystem will eventually recover to its initial state.[13] We need better ecology teaching, including teaching that takes place in the natural environment, on such things as field trips.

Reproduction and Sex

There is a large literature about students' misconceptions concerning reproduction and sex.[14] As might be expected, misconceptions are more likely to be reported about human reproduction than about reproduction in other species,[15] though Schussler[16] showed that children's books about plants not infrequently contain inaccuracies about plant reproduction. Many students have problems in relating the time of conception in humans to the condition of the uterine lining and do not appreciate the connections between menstruation, ovulation and the likelihood of a fertilised egg implanting.[17] Misconceptions about contraception are also widespread[18] and there is a large literature about misconceptions with regards to methods of transmission of HIV.[19]

The term 'misconceptions' may fail to convey all the ways in which scientifically incorrect views are held about a topic, and this is perhaps particularly the case in regards to reproduction and sex. Often, students, especially young students, are simply ignorant — and this is not something for which they can be blamed. As any biologist knows, the diversity of reproduction in the natural world is immense; it is hardly the fault of students if they have been taught very little of this diversity, or even of the specifics of human reproduction. Moreover, when it comes to human sexuality, misconceptions or ignorance are sometimes fuelled by prejudice, for example in regards to sexual orientation and behaviours.

One advantage of teaching about reproduction and sex is that student interest is often high. However, there are a number of difficulties in teaching about these topics. One difficulty, especially with regards to human reproduction and associated topics within sex education, is that teachers may feel embarrassed. The solution to this is better teacher education, whether during their initial teacher training or subsequent professional development, and practice. Support from senior management within schools is important too.

A related difficulty is that in many countries teachers receive little or no training about sex education. Biology teachers, of course, are nearly always trained to teach about reproduction, both in humans and in other taxa. However, the knowledge and skills to teach sex education well are not the same as the knowledge and skills to teach reproduction well. In particular, good sex education makes use of a range of pedagogical approaches (including role plays and debates) that are less often used in biology teaching. Then there is the fact that teaching about sex education raises issues to do with values to a greater extent than when teaching biology. In schools where religious values are important, attitudes towards sex education may be conservative, for example with regards to homosexuality and abortion. Teachers need to be skilled when teaching sex education in such situations.[20] For many teachers, the easier and more pragmatic option may be stick to the biological facts, and this can disadvantage students.

Children's Understanding of Evolution

Evolution is widely seen as the central, key, unifying framework of biology. Yet many school-aged students and adults understand relatively little of the theory of evolution.[21,22] It is almost universally agreed by biologists that natural selection is the most important factor that drives evolution and that therefore three principles are necessary and sufficient to explain almost all evolutionary change: (1) generation of variation, (2) heritability of variation, and (3) differential survival and/or reproduction of individuals with differing heritable traits. However, evolutionary change is still poorly understood by students, throughout their time in education, by science teachers and by the general public.[23–25]

Frequent misconceptions of students are apparent in Lamarckian and teleological explanations of the mechanism of evolution. In Lamarckianism, the features that an individual acquires during its lifetime (iconically, the arm muscles of a village blacksmith) are more likely to be passed onto its offspring. This notion is related to the widespread teleological belief that sees changes as being goal-oriented; new features develop because they are advantageous. Another misconception relates to the distinction between the individual and the population level. The mechanism of selection affects the individual and its interdependency with the environment. Genetic variability leads to different phenotypes and the individuals within a population often show small differences in morphology, physiology and behaviour. Often this level of variation is not appreciated by students. But without such variation, the chances of survival and reproduction are the same for all individuals in the population. Very often, students do not understand what is meant by 'population' or its significance in the context of evolution. This can lead to the misconception that evolutionary adaptation occurs at the level of individuals. Perhaps unsurprisingly, school students also typically have almost no understanding of the process of speciation, even though the fundamentals of the process were correctly described by Darwin in his 1859 classic *On the Origin of Species by Means of Natural Selection.*[26]

Rather than labouring the point that evolution is not well understand, one can more profitably ask why this is the case. It is difficult to compare understanding in one area of biology with another, but it does seem as though evolution understanding is particularly poor. There are two main classes of reasons for this. One is to do with the cognitive difficulties of the theory. Even if we simply focus on natural selection, this is a challenging concept for learners. It requires powers of abstract reasoning and there are a number of steps in the argument, each of which needs to be understood if the overarching concept of natural selection itself is to be understood. And then, of course, there is so much more to evolution than the theory of natural section. For a start, an appreciation of 'Deep Time' is needed — and this is itself a difficult concept for many students, one that can literally be unimaginable. In addition, a learner needs to have an appreciation of the different sorts of competition between organisms, including such things as intraspecific competition for nutrients or mates, whereas many learners think only of interspecific competition, such as in predator–prey relationships.

The second class of reasons why the understanding of evolution is poor is to do with the affective issues that the theory raises for some learners. The most straightforward reading of the scriptures of a number of the world's religions seems to argue against the core evolutionary notion that all organisms are related through descent (vertical transmission) as well as, as we now know, through the horizontal transmission of genetic material. In the case of the Judaeo–Christian scriptures, but not in the Qur'an, there is even an apparent timescale for the early history of life (some of the 'six days' in *Genesis*) which runs completely counter to the aeons that evolution entails — life usually being thought to have evolved on our planet over some 3.8 billion years, timescales that differ by a factor of more than a hundred thousand million. Most theologians have long argued that the scriptures should not be read in this literal way. Nevertheless, many religious believers do so read them. Unsurprisingly, faced with a choice between believing what they are told at home and sometimes in their places of worship versus what they are told in their school science lessons, many youngsters ignore or actively reject what they are told at school, thus hampering their learning about evolution.[28]

There is an additional affective reason why evolution may be rejected, though the literature about this is much sparser. That is because some of the key notions of evolution — that the universe may not have some pre-determined aim, that chance has played a major role in our being here, and so on — can cause existential anxieties.[29,30] Faced with these, a learner may feel safer consciously or unconsciously pushing evolutionary ideas to the back of their mind.

Strategies to Help Children to Learn the Scientifically Accepted Account of Evolution

Understanding of evolution is aided by the sorts of pedagogical approaches that are known to work well in other areas of science education and beyond. In particular, there is evidence that the appropriate use of teaching for metacognition, multimodal approaches, argumentation, inquiry-based science education, reinforcement of learning, context-based learning, models, intercultural dialogic approach and object-based learning can all help promote learning about evolution.[31] To this list we can add that teacher expertise is of great importance. In addition, an increasing amount is slowly being learnt about specific teaching approaches that can help with the vari-ous cognitive challenges (outlined above) that evolution education raises.

One feature that is distinctive to evolution education within biology education is that a not inconsiderable number of learners, depending on the country concerned, come from background where at least some aspects of the theory of evolution (aspects of macroevolution rather than microevolution) are actively rejected on the basis of a supposed clash with religion. As is widely acknowledged, it is not easy for teaching to change the views of those who have creationist beliefs.[32] However, research sug-gests that careful and respectful teaching about evolution can lead to students who initially reject the theory of evolution becoming more likely to accept at least some aspects of it. For example, Winslow et al.[33] found that undergraduates who had been raised by their families to believe in creationism could come to accept evolution by evaluating the evidence for it, negotiating the meanings of *Genesis*, recognising evolution as a non-salvation issue and observing their teachers as Christian role models who accepted evolution.

Finally, it is worth mentioning that there is something of a controversy in the literature about what should be the precise aim of teaching evolution to creationists and others who do not accept it.[34-36] A common view is that evolution educators should aim to get such learners to come to *accept* the theory of evolution; another possibility, one that I hold, is that the aim should be to get such learners to *understand* the theory of evolution — leaving it up them whether or not they accept it. In fact, the difference between these two views, while philosophically important, may be smaller than is sometimes supposed as far as classroom practice goes. Teachers can simply do their best to convey the evidence for evolution and ensure an understanding of evolution, while being respectful of their learners.

References

1. Abrahams I and Reiss M. (2012) Evolution. In: Jarvis P with Watts M (eds.) *The Routledge International Handbook of Learning*, pp. 411–418. Routledge, Abingdon.
2. Barlow HB. (1953) Summation and inhibition in the frog's retina. *J Physiol* **119**: 69–88.
3. Goswami U. (2008) *Cognitive Development: The Learning Brain*. Psychology Press, Hove.
4. Kampourakis K and Reiss MJ (eds.). (2018) *Teaching Biology in Schools: Global Research, Issues, and Trends*. Routledge, New York.
5. Carson K, Dawson V and Venville G. (2018) Microbiology. In: Kampourakis K and Reiss MJ (eds.) *Teaching Biology in Schools: Global Research, Issues, and Trends*, pp. 178–191. Routledge, New York.

6. Jones MG and Rua MJ. (2006) Conceptions of germs: Expert to novice understandings of microorganisms. *Electron J Sci Educ* **10**(3): 1–40.

7. Byrne J. (2011) Models of micro-organisms: Children's knowledge and understanding of micro-organisms from 7 to 14 years old. *Int J Sci Educ* **33**(14): 1927–1961.

8. Okada H, Kuhn C, Feillet H and Bach J-F. (2010) The 'hygiene hypothesis' for autoimmune and allergic diseases: An update. *Clin Exp Immunol* **160**(1): 1–9.

9. Korfiatis K. (2018) Ecology. In: Kampourakis K and Reiss MJ (eds.) *Teaching Biology in Schools: Global Research, Issues, and Trends*, pp. 153–163. Routledge, New York.

10. Hokayem H and Gotwals A. (2016) Early elementary students' understanding of complex ecosystems: A learning progression approach. *J Res Sci Teach* **53**: 1524–1545.

11. Lefkaditou A, Korfiatis K and Hovardas T. (2014) Contextualising the teaching and learning of ecology: Historical and philosophical considerations. In: Matthews MR (ed.) *International Handbook of Research in History, Philosophy and Science Teaching*, pp. 523–550. Springer, Dordrecht.

12. Gotwals A and Songer N. (2010) Reasoning up and down a food chain: Using and assessment framework to investigate students' middle knowledge. *Sci Educ* **94**: 259–281.

13. Ergazaki M and Ampatzidis G. (2012) Students' reasoning about the future of disturbed or protected ecosystems and the idea of the 'balance of nature'. *Res Sci Educ* **42**: 511–530.

14. Reiss MJ. (2018) Reproduction and sex education. In: Kampourakis K and Reiss MJ (eds.) *Teaching Biology in Schools: Global Research, Issues, and Trends*, pp. 87–98. Routledge, New York.

15. Nguyen S and Rosengren K. (2004) Parental reports of children's biological knowledge and misconceptions. *Int J Behav Dev* **28**(5): 411–420.

16. Schussler EE. (2008) From flowers to fruits: How children's books represent plant reproduction. *Int J Sci Educ* **30**(12): 1677–1696.

17. Yip DY. (1998) Children's misconceptions on reproduction and implications for teaching. *J Biol Educ* **33**(1): 21–26.

18. Hamani Y, Sciaki-Tamir Y, Deri-Hasid R, *et al.* (2007) Misconceptions about oral contraception pills among adolescents and physicians. *Human Reproduction* **22**(12): 3078–3083.

19. Harvey I and Reiss MJ. (1992) *AIDSFACTS: Educational Material on AIDS for Teachers and Students*, 4th ed. Daniels Publishing, Cambridge.

20. Halstead JM and Reiss MJ. (2003) *Values in Sex Education: From Principles to Practice*. Routledge Falmer, London.

21. Nehm RH. (2018) Evolution. In Kampourakis K and Reiss MJ (eds.) *Teaching Biology in Schools: Global Research, Issues, and Trends*, pp. 164–167. Routledge, New York.

22. Harms U and Reiss MJ (submitted). The present status of evolution education.

23. Shtulman A. (2006) Qualitative differences between naïve and scientific theories of evolution. *Cognitive Psychol* **52**(2): 170–194.

24. Nehm RH, Poole TM, Lyford ME, *et al.* (2009) Does the segregation of evolution in biology textbooks and introductory courses reinforce students' faulty mental models of biology and evolution? *Evol Educ Outreach* **2**: 527–532.

25. Evans EM, Spiegel AN, Gram W, *et al.* (2010) A conceptual guide to Natural History Museum visitors' understanding of evolution. *J Res Sci Teach* **47**(3): 326–353.

26. Darwin C. (1959) *On the Origin of Species by Means of Natural Selection or the Preservation of Favoured Races in the Struggle for Life.* John Murray, London.

27. Reiss MJ and Harms U (submitted) What now for evolution education?

28. Reiss MJ. (2009) The relationship between evolutionary biology and religion. *Evolution* **63**: 1934–1941.

29. Tracy JL, Hart J and Martens JP. (2011) Death and science: The existential underpinnings of belief in intelligent design and discomfort with evolution. *PLoS ONE* **6**(3): e17349. doi:10.1371/journal.pone.0017349.

30. Newall E. (2017). Evolution, insight and truth? *Sch Sci Rev* **99**(367): 61–66.

31. Harms U and Reiss MJ (eds.) (submitted). *Evolution Education Re-considered: Understanding What Works.*

32. Long DE. (2011) *Evolution and Religion in American Education: An Ethnography.* Springer, Dordrecht.

33. Winslow MW, Staver JR and Scharmann LC. (2011) Evolution and personal religious belief: Christian university biology-related majors' search for reconciliation. *J Res Sci Teach* **48**: 1026–1049.

34. Hermann RS. (2008) Evolution as a controversial issue: A review of instructional approaches. *Sci Educ* **17**: 1011–1032.

35. Reiss MJ. (2011) How should creationism and intelligent design be dealt with in the classroom? *J Philos Educ* **45**: 399–415.

36. Williams JD. (2015) Evolution versus creationism: A matter of acceptance versus belief. *J Biol Educ* **49**: 322–333.

14 Brain-talk and Public Engagement with Neuroscience

Harris Wiseman*

Introduction

In this chapter I will explore two very different discourses about the public perception of neuroscience. The first discourse concerns the popular debate surrounding 'neurohype' and its critics, wherein the public is conceived of as being very interested in brain-talk, saturated with it from various popular sources, and dangerously misled by this brain-talk (call this the 'neurohype/anti-neurohype' discourse). The second discourse, to the contrary, seems to suggest that the general public neither know anything about neuroscience, nor do they care to. In this latter view, the prevalence and impact of brain-talk on the public seems to have been massively exaggerated (call this the 'can't engage, won't engage' discourse), and public engagement with neuroscience is deemed to be floundering. Both of these discourses present some very important truths. I will conclude with an account of what a positive (albeit limited) public engagement with neuroscience might look like — one that moves away from a model of 'knowledge transmission' towards something more reflective, conversational and directed towards

*Research Fellow at Campion Hall, University of Oxford, Brewer St., Oxford OX1 1QS, UK.

examining the larger social significance of scientific debates. The media will be presented as having a crucial role to play in this sort of engagement.

Neurohype and Anti-Neurohype

I will begin with the 'neurohype/anti-neurohype' discourse mentioned above. This discourse has two contrary modes of expression. The first mode is represented in an avalanche of popular neuroscience books, news articles, academic papers, internet videos, all of which take brain-talk very seriously, seeing it as offering important truths about humans and society. The other mode is a considerable body of work which exists to debunk the over-reach of brain-talk, its rhetorical misuse and misapplication in shaping social policy. This anti-neurohype discourse often makes the claim that such popular 'neurobabble' not only misleads the public, which it does by attempting to explain all facets of existence in neurological terms, but is also quite a dangerous phenomenon, giving a scientific veneer to ideological outlooks, and being used to defend entrenched social inequalities. The belief is that brain-talk proffers a reductive discourse that threatens to belittle our sense of self and responsibility. It fools us with a false understanding of the brain, painting a damaging image of ourselves as marionettes, determined by the structure of our brains and the balance of chemicals therein (so-called neurodeterminism).

A central conviction in the neurohype discourse is that low-grade presentation of neuroscience information has saturated public discourse. Steven Poole, in describing this supposed 'intellectual pestilence', presents the situation thus:

> Shop shelves groan with books purporting to explain, through snazzy brain-imaging studies, not only how thoughts and emotions function, but how politics and religion work, and what the correct answers are to age-old philosophical controversies ... This is the plague of neuroscientism — aka neurobabble, neurobollocks, or neurotrash — and it's everywhere.[1]

The more sober account presented by social scientists O'Connor, Rees and Joffe agree with the basic substance of that claim. They write:

> Since the 'Decade of the Brain,' the field of neuroscience has expanded dramatically, tackling increasingly complex topics with profound social and policy implications ... Neuroscience is now firmly rooted as a basic reference point within the public sphere, drawn into discussion of diverse issues such as antisocial behaviour, economic decisions, substance abuse, and education.[2] (p. 220)

Currently, plenty of money is being directed at neuroscience research, as the USA, the EU and China carry through ambitious national neuro-projects, which are of a similar scale to the previous Human Genome Project, according to Francis Collins.[3] Obama's BRAIN Initiative (Brain Research through Advancing Innovative Neurotechologies) committed itself to spending 100 million dollars just in 2014 for its decade-long project, which is interested in 'revolutionizing our understanding of the human brain ... by accelerating the development and application of innovative technologies ... To produce a revolutionary new dynamic picture of the brain ... seeking to treat, cure, and even prevent brain disorders'.[4] Similarly, the EU launched its decade-long Human Brain Project (HBP) aimed at creating a large-scale computer simulation of the brain, and is committing 50 million euros per year to funding related projects. As the MIT Technology Review commented: 'the three-pound mass between our ears is the next great frontier for science'.[3]

It is important to grasp that neuroscientists themselves debate the terms of the projects.[a] Neuroscience is a place for debate, it is not a unified field directed by a singular consensus. Neuroscience's own practitioners emphasise that, so far from having all the answers, neuroscience is still nascent, and that its own metaphors (e.g. the brain as a computer), goals and methods are constantly under critical scrutiny by

---

[a]This point is particularly significant for the coming public engagement section of this chapter — which emphasises communicating *debates* in neuroscience more than the false sense of confidence about its non-medical findings.

neuroscience practitioners themselves. This more modest reality needs to be kept in mind throughout the neurohype debate, whenever hearing big claims presented in the media, as well as when thinking about how governmental policymakers apply neuroscience for directing social policy.

To illustrate such debate, both the USA and EU neuro-projects have been criticised as being premature, both in terms of the technology required for actually carrying out the projects, as well as the lack of the basic knowledge of the brain required to even get started (particularly the EU project). Above all, there is debate amongst neuroscientists themselves as to the reductive terms of the discourse. The MIT Review suggests:

> One reason the US project doesn't satisfy all neuroscientists is that it strongly embraces the idea that there are 'circuits' in the brain, or that one neuron would excite another, and so on, leading to behaviors. While that certainly happens, some say that the circuit analogy is an antiquated anatomical notion insufficient to explain how the 86 billion neurons in the brain actually operate.[3]

Likewise, with the EU project, its technical goals (above all, creating a giant computer simulation of the brain) are considered by some to be ill-conceived, as even its own advocates admit not knowing enough about the brain at the outset to start making credible predictive simulations. The suggestion is that the overall project is likely to fail.[5]

Looking beyond these national projects, thinkers such as Nikolas Rose suggest that brain-talk is virtually omnipresent. Indeed, Rose has gone so far as to suggest that the EU and USA have become 'psychopharmacological societies' where the modification of thought, mood and conduct through pharmacological means have become routine[6] (p. 46). In Rose's view, the general public have become 'neurochemical selves'[6] (p. 46). Rose points out how numerous developments in pharmacology, neuroscience and genomics have transcended the medical and psychiatric domains, and invaded our daily lives and daily thinking. The idea is that our capacities, attention, energy, mood, levels of aggression, and empathy are now routinely conceived of in neurological terms, by the public and professionals across all the domains that have drawn on such neuroscientific findings. As such, neurobiology is no longer just a matter for science professionals, but part of the very language ordinary people use to describe themselves and define their ailments and prospects. Brain-talk, so the suggestion goes, has become the dominant explanatory mode for understanding our basic human experience.

Rose, a very popular anti-neurohype writer, has obviously exaggerated things a great deal, and it seems that some popular anti-neurohype writers are every bit as susceptible to over-hyping their own claims. The neuro-chemical society Rose describes is not recognisable, and his suggestion that members of the public have somehow been reduced to 'neurochemical selves', thinking of themselves as brains to be manipulated, simply does not ring true. Even so, there is some truth in Rose's concerns. At the very least, the use of psychopharmaceuticals has increased in both the UK and in the USA (particularly in the USA where antidepressant use has soared by 65% in the 15 years from 2001–2016;[7] and in 2016, 1 in 6 Americans were being prescribed anti-depressants).[8] Diagnostic criteria which allow for the medical use of such drugs have broadened dramatically; the FDA has ruled that such drugs can be used 'off-label' (i.e. for ailments they were not designed for), and in the USA TV advertisement for the use of such drugs is not only allowed but is big business.[9] Recreational and work-related use of stimulants has certainly increased (ranging from military pilots, to truck drivers, to office workers seeking simply to get a competitive edge on their office colleagues),[10] though exact figures for illicit use are hard to come by.

A balanced view of the extent of neurohype needs to be provided, one which takes account of the genuinely dangerous ways in which such brain-talk has been widely misused, though without falling prey to the wild exaggeration that we sometimes find articulated. Brain-talk has indeed infiltrated the way many institutions and governments defend their policies,[11] and the media all too often present a very crude use of such neuro-reductive language for 'explaining' crime, describing racism as 'hardwired', evil as a neurological disease, depression as a 'chemical imbalance' and faithfulness as hormonal, and explaining actions of rioters and killers in terms of neurochemicals and hormones such as serotonin, dopamine, oxytocin, cortisol, testosterone, and so on.[9,12] As we shall see in the next section, it is debateable how far such discourse genuinely shapes public perceptions. Moreover, I suggest that one should not start with the assumption that the public are quite so gullible as some neurohype critics seem to take for granted — particularly in light of the fact that the anti-neurohype discourse is just as marketable as the hype it rails against. Likewise, I suggest that the media are content to sensationalise *both* sides of a neuroscience debate (both neuro-determinism and neuroplasticity angles) as and when it is convenient.

I will continue a little further with this notion that neuroscience is being overburdened, and being used to explain phenomena far beyond its remit, in particular, social phenomena which are not best described in neural terms to begin with, but which are being almost wholly reduced to neural terms — with only the most superficial lip service being paid to any other potentially relevant factors.

A serious example of this can be seen in the NIH's 2018 advice on alcoholism and addiction. The NIH states explicitly that alcoholism is a brain disease, and can therefore be treated as a brain disease, as a simple malfunctioning of the brain's mechanisms. Commenting on this, Nora Volkow, Director of the National Institute on Drug Abuse, writes:

> Over the past three decades, a scientific consensus has emerged that addiction is a chronic but treatable medical condition involving changes to circuits involved in reward, stress, and self-control; this has helped researchers identify neurobiological abnormalities that can be targeted with therapeutic intervention. … Informed Americans no longer view addiction as a moral failing, and more and more policymakers are recognising that punishment is an ineffective and inappropriate tool for addressing a person's drug problems. Treatment is what is needed.[13]

In a well-balanced assessment, Volkow goes on to suggest that this view might be a step back from the non-reductive biopsychosocial model which currently dominates the USA's addiction treatment thinking, a paradigm 'which recognises the complex interactions between biology, behaviour, and environment'.[13] Even though Volkow observes that 'the underlying concept of substance abuse as a brain disease continues to be questioned, perhaps because the aberrant, impulsive, and compulsive behaviours that are characteristic of addiction have not been clearly tied to neurobiology'[14] (p. 363), she does still 'conclude that neuroscience continues to support the brain disease model of addiction'[14] (p. 363). The sort of neuro-primacy which presents addiction as a brain disease is not some media creation or self-help product. The NIH invests over 37 billion dollars annually into research, and is one of the foremost medical research institutes in the world.[15]

As Volkow notes,[14] labelling alcoholism a 'brain disease' might be a worrying mischaracterisation of a predominantly social issue involving poverty, despair, and/or social attitudes towards the consumption of alcohol. And, Volkow may be right or wrong that addiction is 'not a moral failing', or that punishment does not work. However, while focusing on neurological treatments is a perfectly valid course for a medical research institute such as NIH, defining the problem in such terms allows other state institutions to ignore the underlying social issues with which the neurological 'malfunction' is in a very complicated relationship. It is that *relationship* that needs to be explored, something which touches on society, personal responsibility and biology all connected together.

Neuroscience must be constantly praised for its immense medical value, and is so praised by neurohype critics. Again, the problem comes when this legitimate source of inquiry is misapplied to give causal explanations for things outside its remit. This is considered most illegitimate when used ideologically, politically, or to defend culturally generated divisions. In this, neuroscience is not at all unique — biology generally (and, at present, ecology) are very much being misused as political weapons in so-called 'culture wars'[16] (p. 1). Shi et al. write: 'Passionate disagreements about climate change, stem cell research and evolution raise concerns that science has become a new battlefield'. Shi et al.'s unsurprising finding is that liberals and conservatives are interested in different kinds of science (those, in short, which bolster their own particular views). Science presentation can become a way of propping up whatever one believes in, and the authority of science is taken as a powerful ally. Climate change is particularly divisive in the USA, as Hamilton 2010 writes:

> U.S. public opinion regarding climate change has become increasingly polarized in recent years, as partisan think tanks and others worked to recast an originally scientific topic into a political wedge issue. Nominally 'scientific' arguments against taking anthropogenic climate change seriously have been publicized to reach informed but ideologically receptive audiences … Narrowcast media, including the many websites devoted to discrediting climate-change concerns, provide ideal conduits for channeling contrarian arguments to an audience predisposed to believe and electronically spread them further.[17] (p. 231)

While there will always be some interpretability in scientific theories and discoveries, having a motivation to engage with science which sees it purely as a vehicle for propping up one's already existing convictions is an ever-present temptation — a temptation to which, unfortunately, some scientists themselves, philosophers of science, scholars, media, politicians and certain members of the public have fallen prey.

Hamilton and Shi et al.'s insights suggest that there are 'ideologically receptive' audiences for certain kinds of findings. While it would certainly be too crude to say that the public only hear what they want to hear, it cannot be denied that groups can be receptive to science that fits conveniently with what they already happen to believe. This reality has important implications for the transmission of neuroscience. As O'Connor, Rees and Joffe 2012 write:

> … scientific information is rarely transplanted intact into the public domain. As science penetrates the public sphere, it enters a dense network of cultural meanings and worldviews and is understood through the prism they provide. The cultural context determines which aspects of science travel into public

consciousness: knowledge that resonates with prevailing social concerns is selectively 'taken up' in public dialogue.[2] (p. 220)

There are certainly cases where neuroscience is misused in social policy. For example, Broer and Pickersgill[18] (p. 54) analyse the use of neuroscience in British social policy. They find that:

> Concepts and findings 'translated' from neuroscientific research are finding their way into UK health and social policy discourse. Critical scholars have begun to analyse how policies tend to 'misuse' the neurosciences and, further, how these discourses produce unwarranted and individualising effects, rooted in middle-class values and inducing guilt and anxiety … UK policy reports … employ neuroscientific concepts and consequently (re)define responsibility [to] support a particular imaginary of citizenship and the role of the state.[18] (p. 54)

Yet, in Chapter 7, I argued that courts of law have not been as swayed by neuroscience findings as one might have feared. Neurophilosophical claims that there is no free will (see below), and that jurisprudence needs to be altered to take on board a radically different view of human responsibility have been treated with scepticism. Likewise, I am not convinced that institutions are always naively going to take on board whatever bit of neuroscience they are fed, although, again, there is always a disposition to accept scientific presentations that happen to correspond to what one already believes. Broer and Pickersgill are certainly correct that one needs to be mindful of how far the neuroscientific idiom can justifiably be applied in driving public policy, but it needs to be kept constantly in mind that the neurohype discourse is perfectly well-matched and constantly hounded by neurohype critics debunking the constant overreach, and articulating the dangers of neuroreductive discourse.

Neuroscience? … It's Not for Me

A 'can't engage, won't engage' perspective on the transmission of neuroscience can be construed from sources which seem to suggest that the public are somewhat more disaffected in relation to neuroscience information than suggested above, and in many ways that the public do not feel up to the task of comprehending such science anyway. Above all, this lack of public engagement is a two-way street where experts are deemed to be failing to engage the public successfully, but this is because of a general lack of interest in engagement on both sides. While there has been some recent success in fostering some programs for public engagement, outreach programs and inviting the public into laboratories, such conversation is extremely limited and germinal at best. With public engagement it does 'take two to tango', and where there is no willingness with both parties to attempt to engage with each other, it should not be tremendously surprising that such engagement is not taken up.

It was suggested above that neuroscience (and science presentation generally) is only selectively taken up. Such information tends to be limited to hot topics, and is filtered through a dense network of cultural meanings and worldviews[2] (p. 220). However, later work by O'Connor and Joffe[19,20] highlights how disaffected the public seem to be towards the neuroscience information that is presented to them. O'Connor et al.'s studies suggest something very important about the attitudes of the public towards neuroscience and the scientists who carry out such work. Their research articulates very clearly, with UK audiences at least,[b] the prevalence of a rather sharp sense of alienation amongst the public with respect to the world of science. O'Connor and Joffe[19] write:

> The most dominant mode of relating to science — and thus to neuroscience — was dissociation. 'Science' was positioned as a decidedly separate social milieu in which there was no question of self-participation. The designation of a stimulus as scientific elicited an immediate, patterned response of disengagement from the object in question ... For much of the sample, the domain of science was incontrovertibly 'other,' involving an entirely unfamiliar and 'completely alien' set of understandings, aims,

---

[b]The finding is borne out with American audiences too.[21]

and abilities …. The ability or inclination to engage … was seen to hinge on what 'type' of person one was — namely, whether one was 'scientific' and 'academic.' Respondents avoided brain information because they self-identified as nonscientific, thereby designating the brain beyond their sphere of relevance, interest, and competence … Indeed, some interviewees articulated a sense of discomfort in directly contemplating their brain, experiencing this as cognitively or existentially jarring.[19] (pp. 630, 641)

This is a telling finding. Evidently, a core part of the alienation here is the perception of *identity* differences. It is clear that the word 'scientist' is perceived to represent a distinct identity, one entirely separate from any identity the public associate with themselves. This term 'scientist' seems to evoke in the imagination, qualities, activities and subject matters wholly other than that which the public adopt in their daily lives. Not only is there a stark sense of 'us' and 'them', but this disconnect is concurrent with an active 'pushing away' of science information. As soon as the word 'science' is used, or any symbols perceived to be scientific are employed, there is a sort of gut-level repulsion on the part of the listener. It is unsurprising therefore that neuroscience information has not 'meaningfully infiltrated lay thinking', or that 'Respondents consigned brain knowledge to the "other world" of science … seen as a decidedly separate social milieu'[19] (p. 617).

If the public do not identify themselves in any way as having something meaningful to say regarding science, or worse, as not even being intellectually able to comprehend what scientists are saying, no real engagement with such science can be expected. Note how starkly at odds this picture is with Rose's diagnosis that society and individuals have become so dominated by brain-talk that we have become 'neurochemical selves'. It may be that, at times, brain-talk is prevalent in the news stream. But, if anything, any over-abundance of such brain-talk just causes some persons to tune out. As one interviewee put it: 'I might have seen it on the news or something, you know … But because they probably mentioned the word 'science,' or 'We're going to go now to our science correspondent

Mr. Lala,' that's probably when I go, okay, it's time for me to make a cup of tea. (Male, tabloid reader, 38–57 age-group)'.[19] (p. 630)

Compounding the difficulties here though, and sustaining this sharp identity distinction, has to be the exaggerated nature of the hierarchical divide between scientists and the public. Herein, science is considered to be 'the exclusive preserve of an intellectual elite'[19] (p. 625):

> The perceived complexity of the relevant knowledge precluded lay participation: unfamiliar, dense, and technical language flagged scientific content as 'not for me.' ... Employing vocabulary such as 'extraordinary,' 'noble,' and 'special,' descriptions of scientists were tinged with idealisation and even deification. [One] participant believed that scientists could be trusted precisely because they are 'more than you.' ... While some expressed antipathy toward science, for others science's distanced position fostered an image of admirable beings who conducted work that outstripped the capacities of normal minds.[19] (pp. 625, 631–3)

Again, the language used here is very telling. The scientist is not considered 'normal', indeed is believed to have supernormal capabilities. The response is *demoralisation* — and this is extremely destructive if one is hoping for positive public engagement. One might suggest that the response of being demoralised by science information, and the withdrawal that it creates, is a form of learned helplessness. Using such a term might be helpful in reconceiving the task of encouraging public engagement. This sense of being 'disabled' with respect to comprehending science would need to be addressed. However, one must also wonder whether this sense of inferiority before science is to some extent a convenient way of putting scientists on so high a pedestal that one feels legitimately excused for not meaningfully engaging. One of the most serious consequences of such a withdrawal from science is a relinquishing of *responsibility* regarding the evaluation and direction of science. If science information is 'not for me', or if I am not even capable of grasping it, there can be no attitude of responsibility towards it. So, demoralisation, or a simple unwillingness to take on this responsibility, produces a shift in agency away from the public into the hands of what are perceived to be superhuman experts (or at least, the policymakers these experts guide).

If members of the public in no way regard themselves as even being able to comprehend, let alone contribute constructively to, the evaluation and direction of science, how can there be any reasonable expectation of a responsible public attitude towards science? While there may be plenty of legitimate reasons why the public might not want to engage with science, the belief that scientists are superhuman experts is not one of them. Neither is the belief that the public are in no way capable of grasping the contours, directions and moral implications of current debates about the progress of science and its application. For example, one might look at the current debate surrounding AI research, whose advancement relies in no small measure on the utilisation of massive datastreams composed of the general public's private personal information. It does not take a superhuman intellectual to grasp the significance of the debate, nor to have a meaningful opinion on it. Less stringent privacy laws and deregulation enhance the speed of AI development, so there must be a debate about what trade-off is acceptable here. This is precisely the sort of debate that the general public need to get involved in, because it is the public's data that are used as grist for enhancing AI's predictive and pattern-recognition powers.

Neuroscience and the Public: Us versus Them

In terms of engaging with neuroscience, one of the big obstacles in bringing brain information into public discussion is the fact that neuroscientific assumptions about the mind and brain are very different to those routinely employed by the public[22] (p. 108). Unsurprisingly, mind and brain are approached very differently in the laboratory and everyday life. If the foundational assumptions applied by neuroscientists and the public are so very different, then the difficulties of translating between them is heightened — particularly, as some neurophilosophers have it, because many persons are committed to certain views on freedom and personal identity which current neuroscience seems to contradict. The notion of 'eliminative materialism' in neuroscience takes the point that there is an abyss between public understanding of the brain and 'the actual science' very seriously.

Patricia Churchland (along with her husband), pioneered the notion of 'eliminative materialism' in neuroscience. The 'eliminative' aspect means that, as we discover more and more about the brain, previous categories that are still used in common parlance are found to be inappropriate, or false presentations of what is 'really going on' in the brain.[23] Churchland gives 'the will' as an example of something we are used to talking about as if it were a distinct power or faculty, and points out that the notion of 'the unified self' is but a helpful shorthand — one that serves us in everyday speech, but which is really just a useful illusion. These popular illusions need to be removed from scientific discourse because they do not fit with what neuroscientists are finding out about the brain.

This elimination of everyday thinking from neuroscientific theorising only serves to create a larger and larger wedge between public and neuroscientific thought worlds. As Churchland emphasises, the illusions of consciousness are to be eliminated, *not* from daily talk (where they are helpful and necessary for getting on with living), but from scientific discourse only. There are to be two distinct thought worlds, it would seem, the lived world where illusions reign, and the scientific world, where truths about material and mechanistic processes reign.

Elizabeth Valentine explicates the extent to which neuroscience and psychology are increasingly diverging from popular views of ideas of self, will and freedom. Valentine writes: 'There is now overwhelming evidence to invalidate the concept of consciousness as a transparent, unified, coherent system that plays a role in the control of behaviour'[24] (p. 57), pointing out how empirical work on 'blind sight' phenomena, subliminal and implicit perception, and the machinations of implicit memory seem to undermine this idea of consciousness as the seat of executive control. Valentine continues at great length:

> Discrimination, perception, memory, thinking, judgment and problem solving can all occur outside conscious awareness. Thus, they cannot be criterial of consciousness ... Consciousness appears to be tangentially involved at a late stage ... Nor does consciousness control behaviour, which is determined non-consciously. Consciousness may be unified and coherent within itself but it is based on an extremely small portion of the total picture. We are systematically misled by the limited perspective of consciousness ... we falsely believe that conscious processes determine behaviour ... Our sense of continuity and control is illusory.[24] (pp. 61–62)

The emphasis is clear: 'we aren't consciously aware of events at the time they happen, but we think we are'[24] (p. 62).[c]

It may very well be that Churchland, Valentine and their many advocates do not represent the views of all neuroscientists (though the public would not know that from the way the popular neuroscience debate is framed). Even so, the following point needs to be made: it is not just the public that construct this alienating division between scientific and public thought worlds. Many scientists are, some without realising the implications of their speech, entirely complicit in constructing and sustaining this divide. The neuroscientific discourse on mind and brain is extremely clear on this: there is public thinking, and that is ridden with illusions; then there is science thinking, and that is reality and truth. There is a line between the two thought worlds, and that line is sharp and clear. Lay thinking is deluded, science thinking is not. Given that these are the terms of the popular neuroscience debate on mind, brain and freedom, there can be

---

[c]It is to be noted that Valentine is not an eliminativist inasmuch as she rejects the notions that conscious experience tells us nothing scientifically useful. Valentine argues that 1st and 3rd person perspectives are both absolutely necessary for studying the relation of consciousness to the brain.

no surprise that there is a sense of alienation between public and scientists in the neuroscientific domain.[d]

This is an unnecessarily impoverished relationship. Instead, the sense that the public do have an active stake in the direction of science needs to be fostered. Science affects everyone. The larger social implications of neuroscience research (and science generally) need to be made explicit. Much science relates directly to the human person, environments on this planet, and the various food that we consume, all of which issues touch everyone's daily lives at every point during the day. It is important, therefore, that a sense of this intimately interwoven relationship between science and our everyday lives be cultivated. Everyone needs to have at least some sense that we have a stake in how science is to proceed, as well as a sense that we might have something valuable to say on the subject, as well as the sense that our voices will be heard.

Public Engagement with Neuroscience

According to the UK Government's *Public Attitudes to Science* Report conducted in 2014 (PAS2014)[26]: 'the UK public overwhelmingly thinks that science is important and takes an interest in it. There has been a gradual long-term increase in support, with the public appearing much more interested in science today than they were in 2000 and before'.[26] But what does PAS2014 really tell us? On the one hand, it says that the public feel positively towards science. On the other hand, it discovered that actual scientific knowledge amongst the public is quite poor, particularly with respect to genetics.[27] I think this report is helpful for enabling us to disentangle a range of ideas. The investigators claim that they wished to 'focus

---

[d] Dennett is particularly patronising when it comes to public comprehension of mind and brain: 'Consciousness is just a bunch of mundane magic tricks the brain plays to give itself the illusion that it is special'.[25] He regards persons who think otherwise as being 'in denial', as sentimentally deluded — the exact words he uses — clinging desperately to illusions that they need to maintain their fragile sense of comfort in the world.

on the social connections people have with science … This is why we poll for public attitudes, not public understanding'.[28] But, the basic fact that PAS2014 reveals is that public sentiment towards science simply does not translate into any comprehension of science, nor any meaningful engagement with it. Currently, science seems to be interesting — from a distance.

PAS2014 has been promoted with an immensely congratulatory feel about its message. People feel positively about science. But, must one not question the value of this sentiment if it does not translate into anything more? As Zadik asserts, there are implications for the poor understanding of science that PAS2014 revealed. When scientists and journalists use scientific terms, it cannot be assumed that the reader understands what is written.[29] Perhaps the news presenters do not understand much better either. Zadik asks: what good is a news article on the human genome if the public do not even know what a genome is? One might suggest that this 'positive regard' towards science is simply a way of washing one's hands, of saying something like: 'I'm happy to just let the scientists get on with it, and I neither need to know nor engage with what is going on'.

Such a view is problematic. Certainly, science continues to do a very good job of mitigating many of our anxieties and problems. It seems that in many countries one can take it almost for granted that food will be available to buy, that healthcare can and will be provided, and that all ailments and illnesses will be curable one day. Unfortunately, the notion of science as offering certainties, the notion that science is on a triumphant campaign to rid us of all our insecurities is unrealistic. Science may have dispensed with some previous anxieties but it has also created some new ones. Cutting edge science rarely has any kind of consensus; scientists are ordinary people, with their own values and beliefs, interests, pressures and temptations. Moreover, some scientific pursuits have graver global consequences than others. Therefore, popular myths about what scientists look like, and what science is, need to be dispelled. The idealised view of science and scientists as superhuman experts only reinforces false stereotypes and sustains the unhelpful sense of alienation between the public and the science community.

Yes, science provides models of the world with explanatory and predictive power, and has been tremendously successful in doing so. But, it needs to be made abundantly clear that science is not a magic wand, that offering certainties is not something it can do, and that science raises global issues that reach into all our lives.[e] A completely laissez-faire approach to science, one based on a mindless 'positive regard' with no basic comprehension of science or its implications, cannot be a fruitful way forward.

What Would a Positive Public Engagement Look Like?

There are certainly many impediments to public engagement with science. But, what would a meaningful engagement look like? Zimmerman and Racine furnish us with a very helpful starting point which attempts to capture the present range of interactions between scientific research and the public[30]:

(a) public understanding
(b) knowledge translation
(c) public participation
(d) social outcomes
(e) dual use

According to Zimmerman and Racine, 'coverage of these topics is sparse and inconsistent in mainstream policies and guidelines'[30] (p. 27). Even so, these five categories are useful for clarifying how engagement can be profitably construed. Above all, what we need is something moving

---

[e]References to the false perception of science being able to offer certainties and technofixes for assuaging all human anxieties are taken from conversations had with Ottoline Leyser at the International Society for Science and Religion's 'New Biology' Symposium, Cambridge, September 2016.

far beyond the dominant didactic model. Science communication is too often reduced to the mere transmission of 'science facts', as if from above, for public absorption. This is a passive receptor relationship, and such 'communication' is more of a monologue than a conversation. As Bobby Heagerty, Director at the American Brain Coalition, and very critical of the current mode of neuroscientific engagement with the public, puts it: 'Talking to the public (some thinkers claim "talking at the public" is more accurate) — dissemination — is not the same thing as *engaging* the public. Dissemination goes one way; as information flows from those who are knowledgeable to those who are not. We need to change the language as well as the approach'[21] (p. 4483).[f]

In Zimmerman and Racine's presentation, we see that knowledge translation is but one aspect of public understanding. The knowledge to be translated has a broader context — above all, the social and global outcomes of science. While there is nothing wrong with absorbing science documentaries, that is not the point of engagement. It is the equivalent of donating a few pounds to charity and thinking one has solved world hunger. Engagement needs to be aimed at providing some sort of reflection or critical oversight into the manner in which science is progressing — engaging with particular debates and issues. Rather than just remarking upon where a particular scientific line of research is heading, an element of 'ought' needs to be involved. In this way, science is to be treated in a similar way to politics. Science does have, and always has had, political implications. The public should understand that the march of science involves debates about the ethical limits of science, and what trade-offs are required in pursuing certain scientific lines of research (e.g. energy consumption brings concerns over pollution and global warming, the safety and viability of nuclear generators; gene editing raises questions of designer babies, and so on).

---

[f]See Heagerty[21] for a wealth of recommendations on how to stimulate a rich dialogical public engagement with neuroscience and why such engagement is much needed.

It is in this regard that the most degraded and sensationalist portions of media coverage of science are so destructive. As with politics, the media are a core channel through which the democratic function is carried out. Insofar as science has political, ethical and social ramifications that impact national and international relations, debates in science need to be presented in a responsible manner. Manipulation of the media and ideologically-based reporting bent on generating popular uprisings can be as dangerous in science as with politics. The stirring of public disapprobation in the GM food debate created a storm of antipathy, and represents a perfect example of how public engagement can be woefully mishandled. Safety reassurance was not communicated well by the scientific community, nor did the public hysteria facilitate healthy debate. The media, generating a polarised debate and stirring up both sides, did not help either. There was little sense on any side that *the very ability to have a responsible debate* is at least as crucial as the issues being debated.

A cynic might say that this sort of debacle is a good reason *not* to have public engagement at all, and to simply leave things to the experts, to let the scientific community police itself (and, in all fairness, in many cases scientists can be relied on to be self-critical — in the gene editing debate, for example, there are plenty of scientific cautionary voices and sound debate over limits has prevailed). However, the fact is that while scientists certainly do debate boundaries and ethics amongst themselves, there are financial and institutional pressures at play, as well as furious international competition for prestige, and such factors constantly threaten to undermine responsible action on the part of the science community. And, this is why it is essential that these issues be presented to the public, and that the public be part of the conversation. If the academy is constantly under threat from market forces, as well as political and institutional pressures, there needs to be a level of oversight that goes beyond the academy and governmental sources — even if the risk of mishandling public engagement exists.

Public engagement must combine a digest of knowledge with actual understanding of the larger issues that such research relates to. I would suggest that a large part of the burden for responsible communication needs to fall on the media's shoulders. Not all scientists are PR people, nor should they be. One must be careful, in calling for public engagement, not to overburden the scientists whom may have not have aptitude, willingness or time to engage in this manner. I think a more careful and limited account of the expectations to be put on scientists' shoulders would be more beneficial. Indeed, while public engagement is important, there is a danger of taking such engagement too far. The distortions created by the need to make one's science seem as if of historical importance are well known. An unfortunate mutually exploitative relationship exists between scientists, government grant bodies and the media.[9] Use of buzz words, and, in particular, putting the word 'Big' before whatever one is talking about ('Big biology', 'Big ecology', 'Big neuroscience') tends to be very successful in opening up streams of money. Yet, these sensational aspects all too often do not reflect the actual science being done. This means that the public discourse is going on in a space that has little to do with the actual science. This sort of engagement does not foster truthful communication, in fact it does the exact opposite, obfuscating the reality of what scientists are doing.

It must also be acknowledged that there are numerous factors standing in the way of a broad public engagement with science, and one must be realistic about one's expectations. First, most people are simply too busy to be expected to engage in any overly-demanding concrete manner (the University of Leicester's outreach programs, GENIE,[31] inviting the public into its laboratories to see how science is done, is a wonderful gesture, but this is not the sort of thing one can expect sufficient members of the public to participate in). Secondly, while people express a general positive regard to science when questioned, one might suspect that the amount of time and energy they are really going to devote to engaging with such science is likely to be modest at best. As such, a media-dominant model, one that parallels the media's political function in responsible reporting to sustain

---

[9]Once more, I am indebted to conversations with Ottoline Leyser regarding these realities.

the nation's democratic function, is probably the best way (perhaps the only way) to cultivate a long-term ongoing public relationship with science. The media present the best way to reach a large number of people, quickly, and to stimulate conversation and action. The issue, then, becomes one of fostering a strong sense in media outlets regarding the importance of their role, and a strong sense of responsibility on their part.

Moreover, just as a functioning democracy does not require every single citizen to be informed and active, likewise, the public engagement with science can tolerate those who are not interested, just so long as there is a sufficient base of persons willing and able to read high quality science information, reflect on it, have an opinion on it, and, above all, participate in a conversation.

The hope might then be a recovery of the ability and wish to participate in scientific debates.

There needs to be some meeting of minds wherein differences of values and opinions can be understood and reflected on rather than only having adversarial debates (the neuroscience of free-will public debate is exactly of this polar and gladiatorial kind). Even so, a great deal of work needs to be done in shifting the notion of the public as 'the audience', sat in the stalls, on the other side of the action, into something more conversational. Expecting people to give up a lot of their time to outreach programs is unrealistic. But, high quality media reporting, which raises the issues, the debates, and fosters the kind of engagement just described, can do much more. Mass action, for good and ill, has been the result of well-timed reporting. The media have power. And, just as responsible political reporting is crucial to our democratic function, a similar attitude must be the soundest way forward for a broad and effective public engagement with science.

References

1. Poole S. (2014) Your brain on pseudo-science — the rise of popular neurobollocks. *NewStatesman*. https://www.newstatesman.com/culture/books/2012/09/your-brain-pseudoscience-rise-popular-neurobollocks [6 September 2014].

2. O'Connor C, Rees G and Joffe H. (2012) Neuroscience in the public sphere. *Neuron*. **74**(2): 220–226.

3. Regalado A. (2014) Obama's brain project backs neurotechnology. *Technology Review*. https://www.technologyreview.com/s/531291/obamas-brain-project-backs-neurotechnology/ [30 September 2014].

4. *BRAIN Initiative* Website (2018) https://www.braininitiative.nih.gov/ [18 October 2018].

5. Regalado A. (2014) Neuroscientists object to Europe's human brain project. *Technology Review*. https://www.technologyreview.com/s/528796/neuroscientists-object-to-europes-human-brain-project/ [7 July 2014].

6. Rose N. (2015) Neurochemical selves. *Society* November 43–59.

7. Mundell EJ. (2017) Antidepressant use in U.S. soars by 65 percent in 15 years. *CBS News*. https://www.cbsnews.com/news/antidepressant-use-soars-65-percent-in-15-years/ [16 August 2017].

8. Fox M. (2016) One in 6 Americans take antidepressants, other psychiatric drugs: Study. *NBC News*. https://www.nbcnews.com/health/health-news/one-6-americans-take-antidepressants-other-psychiatric-drugs-n695141 [12 December 2016].

9. Wiseman H. (2014) *The Myth of the Moral Brain — the Limits of Moral Enhancement*. MIT Press, Cambridge, Mass.

10. Pedersen W, Sveinung S and Copes H. (2014) High speed: Amphetamine use in the context of conventional culture. *Deviant Behav* **36**(2): 146.

11. Illes J, Kirschen M and Gabrieli J. (2003) From neuroimaging to neuroethics. *Nat Neurosci* **6**(3): 205.

12. Wiseman H. (2018) The sins of moral enhancement discourse. In: Hauskeller M and Coyne L (eds.) *Moral Enhancement: Critical Perspectives*. Proceedings of the Royal Institute of Philosophy. Cambridge University Press, Cambridge.

13. Volkow N. (2018) What does it mean when we call addition a brain disorder? *National Institute on Drug Abuse*. https://www.drugabuse.gov/about-nida/noras-blog/2018/03/what-does-it-mean-when-we-call-addiction-brain-disorder [23 March 2018].

14. Volkow N, Koob GF and McLellan T. (2016) Neurobiologic advances from the brain disease model of addiction. *New Engl J Med* **374**: 363–371.

15. *National Institutes of Health* (NIH) website https://www.nih.gov/about-nih/what-we-do [18 October 2018].

16. Shi F, Yongren S, Dokshin F, *et al.* (2017) Millions of online book co-purchases reveal partisan differences in the consumption of science. *Nat Hum Behav* **1**(0079): 1–2.

17. Hamilton L. (2010) Education, politics and opinions about climate change evidence for interaction effects. *Climatic Change* **104**(2): 231–242.
18. Broer T and Pickersgill M. (2015) Targeting brains, producing responsibilities: The use of neuroscience within British social policy. *Soc Sci Med* **132**: 54–61.
19. O'Connor C and Joffe H. (2014) Social representations of brain research: Exploring public (dis)engagement with contemporary neuroscience. *Sci Commun* **36**(5): 617–645.
20. O'Connor C and Joffe H. (2015) How the public engages with brain optimization: The media-mind relationship. *Sci Technol Hum Val* **40**(5): 712–743.
21. Heagerty B. (2015) Dissemination does not equal public engagement. *J Neurosci* **35**(11): 4483–4486.
22. Grotzer T. (2011) Public understanding of cognitive neuroscience research findings: Trying to peer beyond enchanted glass. *Mind Brain Educ* **5**(3): 108–114.
23. Churchland P. (2015) How does brain research change our understanding of thinking? *Serious Science*. http://serious-science.org/neurophilosophy-4079 July 6 [18 October 2018].
24. Valentine E. (2014) *Philosophy and History of Psychology — Selected Works of Elizabeth Valentine*. Psychology Press, New York.
25. Dennett D. (2009) Daniel Dennett explains consciousness and free will. *Big Think*. https://bigthink.com/videos/daniel-dennett-explains-consciousness-and-free-will. March 9 [18 October 2018].
26. *Department for Business, Innovation and Skills* (UK government) (2014) PAS201 Report. https://www.gov.uk/government/publications/public-attitudes-to-science-2014 [18 October 2018].
27. Chapman R, Likhanov M, Selita F, *et al.* (2018) New literacy challenge for the twenty-first century: Genetic knowledge is poor even among well educated. *J Community Genet* 1–12.
28. Bell A. (2014) How to read the latest data on public attitudes to science. *The Guardian*, https://www.theguardian.com/science/political-science/2014/mar/14/how-to-read-the-latest-data-on-public-attitudes-to-science [14 March 2014].
29. Zadik D. (2013) The Meaning of 'Genome'. *Wellcome Trust*. https://blog.wellcome.ac.uk/2013/06/28/the-meaning-of-genome/ [18 October 2018].
30. Zimmerman E and Racine E. (2012) Ethical issues in the translation of social neuroscience: A policy analysis of current guidelines for public dialogue in human research. *Account Res* **19**(1): 27–46.
31. GENIE — Genetics Education. *University of Leicester*. https://www2.le.ac.uk/projects/genie [18 October 2018].

15 The Public Understanding of Biology: A Journalist's Perspective

Mark Vernon*

A personal note might be useful. My scientific training extended to a physics undergraduate degree. I've written regularly on science at a popular level for about 20 years, particularly in relation to science and technology, and science and religion. And I've faced the difficulties of writing about biology, and evolution in particular.

For example, one article was published in the Church Times. It commented on the significance of the 'new science' of epigenetics — 'new', as everyone knows, being one of the ways of attracting a reader and editor's eye, like 'sex' and 'chocolate'. It also happens to be an accurate description of the science of epigenetics, I thought. But several letters to the newspaper, from apparently well-informed sources, told me this is wrong. Epigenetics has been around for decades. At the most basic level, the article was deemed by the experts to have failed.

---

*200 Benhill Road, London SE5 7LL, UK.

Evolution is Various

The possibility that everything in biology can be contested lies at the heart the problem journalists face. Without the qualifications of heavily footnoted academic texts, can anything true be said briefly and with clarity?

The challenge comes to the fore particularly in relation to Darwinian theory, the bedrock of modern biology. On the one hand, it is routinely presented as a uniform, unchanging, irrefutable proposition, perhaps in contradistinction to other 'flakier' grand narratives that were also born in the 19th century, such as Marxism and Freudianism. And yet, on the other hand, in its details, Darwinism has been contested from the start. To this day, arguments blaze in relation to the validity of different approaches. The model that dominates the public imagination, which was explained so lucidly in Richard Dawkins' bestseller, The Selfish Gene, is now, I understand, thoroughly contested.

Karl Popper called Darwinism a 'framework for testable scientific theories'[1] (p. 195). One of its testable theories is that life adapts by trial and error, with variances being eliminated when they do not successfully survive. The genius of Charles Darwin, Popper said, was to show that this happens over very long periods of time. It's the essence of evolution as taught by its advocates. It's what Daniel Dennett calls a 'universal acid', and is heralded as being capable of explaining anything.

However, it is not hard to find mainstream, tenured scientists who wouldn't agree, even over the core idea of the length of time that adaptation requires. It is regularly questioned, as in the theory of 'punctuated equilibrium'.

Further, Darwinism attracts theories that can only be inconclusively tested, such as proposals on the origins of life. They have a lesser scientific status partly because the origin of life cannot be observed and so theories about it must rely not on deduction but induction. They are also of questionable scientific value, according to Popper, because any theory of life's origins will involve very low probabilities that morph into high probabilities because of the immensity of evolutionary time. That conversion from low to high, given enough time, undermines the science because it is a principle that enables the generation of explanations for almost anything: a universal acid. (In physics, there is a parallel problem, though there it's recognised as a problem: the Boltzmann brain problem.)

Science is Politics

What this all means in practice is that the communication of evolution inevitably becomes as much a matter of politics as evidence. It's not to say that evolutionary theory is not in some sense right. It is to say that the way it is right is hard to pin down, and that's a problem for journalists. Directly or indirectly they must decide on matters that are extra-scientific, such as the reputations and weight of the scientists they might cite.

For the same reason, culture wars invariably play a massive part in reporting. This is arguably worse in the case of evolutionary biology, than say physics, because to the untrained eye it can easily seem as if the uncertainties of Darwinian theory make way for the unscientific proposals of creationism and intelligent design. Journalists who are perceived to have opened the door to these possibilities will receive quite extraordinary amounts of bile and persecution via websites and social media. Until you have experienced it, it is hard to appreciate just how challenging such attacks are.

In short, there is always an agenda in popular writing about biology, and that conflicts with the image of science as having a vantage from which to judge itself and other forms of knowledge. As the historian of science, Peter Harrison, has pointed out, science can implicitly define itself not as a neutral method for producing knowledge but as successful insofar as it can sustain a worldview that supplants other worldviews, notably religious ones.[2] This is the battleground into which step writers on biology.

Statistics is Hard

There are other tensions and conflicts at play. For example, much of the rhetoric around Darwinian ideas declares its supreme elegance and simplicity, and yet it is also obviously the case that the theory and evidence are often hard to grasp. The reasons for this are partly conceptual. It is easy to perceive how adaptations work over one or two generations. A finch with a beak that can crack nuts with thicker shells will be better able to survive a seasonal drought that causes shells to toughen. But it is hard to visualise how adaptions work over the extensive periods of time required for fish to evolve into humans. To put it another way, biology is a statistical science, natural selection is a population-based mechanism, and statistics is mostly counterintuitive.

A related difficulty, coming to the fore now, is the relationship between top-down and bottom-up explanations. Some of the most interesting developments in modern biology address the environmental context within which evolution occurs, as opposed to focusing solely on the genome. My piece on epigenetics was an attempt to portray this research. However, reductive explanations are much simpler to convey than the fuller, integrated picture. Parts can be grasped imaginatively, whereas wholes often cannot.

Inadequacies of language show up in other ways. There's the fact that whilst evolution does not arise from effort or design, metaphors implying effort and design are routinely used to explain it. The very word 'selection', in natural selection, can't but help imply as much in common usage and other linguistic inconsistencies populate the communication of biology. Keen advocates will describe the science as 'soulful', even 'spiritual', and yet simultaneously insist it is 'mindless'. They will say it is 'beautiful' as well as 'blind', 'inventive' and also 'purposeless'.

To the everyday reader, and I expect to the learned scientist, these contradictions are confusing. They particularly show up in the effort to communicate the biology to wider audiences, and why wouldn't they? As Mary Midgley has pointed out, they expose one of the fundamental challenges for modern biology. It aims to describe life and yet is 'life-blind', as the philosopher puts it.[3] It doesn't account for many of the basic experiences directly known by living creatures, including humans, such as consciousness.

Celebrity Challenge

Given such complications, journalists adopt various strategies to simplify things. A prevalent one is to turn to the celebrity scientist and treat them as the authoritative source.

Richard Dawkins is the best-known case in point when it comes to biology. Research has shown how his prominence, which is as much to do with style and clarity as knowledge and expertise, distorts the communication of the science. One examination focused on the opinions of biologists and physicists in British universities. It asked them about the impact of celebrity scientists on science communication, and concluded that in the public sphere the scientist is asked to be a provocateur as much as an expert.

'Critics, who include both religious and nonreligious scientists, argue that Dawkins misrepresents science and scientists and reject his approach to public engagement', the researchers write.[4] They show that scientists without a particular gift for science communication wanted to 'emphasize promotion of science over the scientist, diplomacy over derision, and dialogue over ideological extremism'. The difficulty is that, for popular media, the personality of the scientist is more important than the science; derision is more arresting than diplomacy, and ideological confrontations draw readers and viewers more effectively than cool dialogue.

Causes and Correlations

It's important to stress that basic misunderstandings of science are also rife amongst journalists. One of the most widespread is the conflation of causes and correlations.

A recent review, published in The Lancet, identified risk factors linked to dementia. It generated headlines about how to avoid dementia by making lifestyle changes. These included such diverse strategies as raising levels of education and addressing midlife hearing loss, as well as tackling obesity and smoking. 'Risk factors' became causes in the reporting, and what was easy to overlook, too, was that the factors themselves actually accounted for only a small fraction of the overall risk. In fact, two-thirds of dementia risk is thought to be 'non-modifiable', The Lancet reported. In a way, that was the real story, but it doesn't make for good headlines.

There are other examples of the problem with causation and not all can be laid at the journalist's door. One that is endemic is the identification of brain activity with everything from memory to happiness. It oversteps the highly controversial identification of mental states with brain states, and is hardly ever challenged.

Genes For

The statistical nature of modern biology presents related problems. It's found particularly in the reporting of genetic research. The public has, as it were, not yet given up on the hope that genes will be discovered for many, if not most, features of human life and many, if not most, causes of human ills. Again, why would they think otherwise, when the expectation was widely propagated by scientists before the turn of the millennium?

The upshot, today, is that the same piece of genetic research can lead to widely disparate stories. Take the publication of evidence from a large study which showed that more than 500 genes are linked to intelligence. One broadsheet newspaper leapt on the findings by commenting that 'intelligence could be measured with a swab of saliva or drop of blood, after scientists showed for the first time that a person's IQ can be predicted just by studying their DNA'.[5] That was in marked contrast to a leading science magazine, which concluded almost exactly the opposite: 'However, even with all these genes, it's still difficult to predict a person's intelligence from their genomes'.[6] It took the considered leader of another newspaper to point out that the notion of genes for intelligence is 'going beyond the evidence'.[7] At least it was pointed out somewhere.

Beware the Reductions

So, if scientists lament the collapse of crucial distinctions between cause and effect, and the relationship between individual cases and population studies, they are themselves guilty of making similar reductions in other domains.

The philosopher, Harris Wiseman, has examined the highlighted hopes for biotechnology, from which is promised human advances such as 'moral enhancement'. An oxytocin spray may make someone more generous. Titrating serotonin may reduce anger, it's said. But this makes no sense to psychologists and philosophers who study generosity and anger and see that these qualities are more than one thing. Generosity, say, may be linked to characteristics as disparate as dutifulness, carelessness, and joy. Causes and correlations are, again, being blurred and journalists aren't only to blame.

Part of the story here is that well publicised promises help secure funding. It's been going on within the scientific community ever since pharmaceutical companies promoted antidepressants on the basis that low mood is caused by 'chemical imbalances'. There is no evidence for this, but it is an easily understood message and has become the received wisdom.

Occasionally, the reduction of complex human phenomena reaches levels of farce. At the time of writing, there was a news story doing the rounds on the ability of killer whales to speak. Serious publications carried discussions of the emerging world of 'cross-species chat'.

Of course, the difference between an animal's grunt and the human animal's deployment of vocal symbols and abstract syntax is vast. To conflate the two is comparing chalk with cheese, coupled to the fact that the origins of language are as little proven as the origins of life. But it is not hard for journalists to find evolutionary biologists prepared to overlook these distinctions and, again, I suspect many are politically not scientifically motivated. The story about the talking orca appealed as it was an opportunity to attack the differences between humans and other animals. That so-called 'myth' has been a widespread target for contemporary biological science partly because it's perceived to be one that religions have upheld.

Peter Harrison's point is moot: some feel that science succeeds when it supersedes religious worldviews.

It's an example of the agendas to which the science is put, and with which journalists wittingly and unwittingly collude. At times, it can seem as if nothing less than the status of biology and science is at stake, which points to one final point: the communication of the limits of biology and awareness of how its great strength, empirical research, is also a weakness, when the data isn't illuminating.

Limits and Promissory Notes

On some occasions, these limits are admitted. An example was the television documentary, The Secret Science of Pop, presented by the evolutionary biologist, Armand Leroi. It is easy to see why the programme was commissioned. Leroi is charismatic on screen, the backtrack of the film could be compiled from numerous catchy songs, and then there's the use of that little word, 'secret'. Science was going to reveal it all. Only, it didn't.

Leroi and his team of experts loaded past hits and chart data into computers, explaining how it was going to be analysed by 'cutting-edge' and 'innovative' artificial intelligence. The promise was that the AI would see what the human creators of chart-topping songs could not. But the number-crunching came up with almost nothing to go on and, at the end of the programme, Leroi confessed to camera that the exercise had been a failure.

It might have been a wise moment for evolutionary biology, a chance to explore what science can and can't achieve. Only, the programme stuck to script, and what Popper called science's 'promissory note'. Leroi insisted that just because the computers aren't powerful enough to penetrate the secrets of pop success today, doesn't mean they won't tomorrow.

It's a human hope. It's the desire for progress, which is another factor that profoundly shapes the reporting of biology. Whether or not it is really part of science is another question entirely.

References

1. Popper K. (1992) *Unended Quest*. Routledge Classics, London.
2. Harrison P. (2015) *The Territories of Science and Religion*. University of Chicago Press, Chicago.
3. Midgley M. (2014) *Are You An Illusion?* Routledge, London.
4. Johnson DR, Ecklund EH, Di D and Matthews KRW. (2016) Responding to Richard: Celebrity and (mis)representation of science. *Public Underst Sci* **27**(5): 535–559.
5. Knapton S. (2018) DNA tests can predict intelligence, scientists show for first time. *The Telegraph*. https://www.telegraph.co.uk/science/2018/03/12/dna-tests-can-predict-intelligence-scientists-show-first-time/ [12 March 2018].
6. *New Scientist*. (2018) Found: More than 500 genes that are linked to intelligence. https://www.newscientist.com/article/2163484-found-more-than-500-genes-that-are-linked-to-intelligence/ [12 March 2018].
7. *The Guardian* editorial. (2018) The Guardian view on intelligence genes: Going beyond the evidence. https://www.theguardian.com/commentisfree/2018/apr/01/the-guardian-view-on-intelligence-genes-going-beyond-the-evidence [1 April 2018].

Afterword

Celia Deane-Drummond*

Anyone reaching the end of this book will be aware not only of the range of topics currently under discussion among biologists today, but also the importance of scientists being willing to present their work in a way that does not loose nuance regarding the essential driving questions and complications that makes the intellectual task of biology so fascinating. Biology as a discipline is promiscuous; at its most creative it does not just keep to its own boundaries, but spreads into other areas of research such as medicine, anthropology or engineering where biological insights are both significant and potentially transformative. I cannot possibly do justice to the full breath of ideas and discussions that this book has covered, but rather offer my own perspective on where I think biology may be moving in the light of my own experience as a biologist and from my own continued fascination with biological concepts as I moved into research that was more deliberately engaging with ethical and theological frameworks. Any creative endeavour takes risks, and, given that I write this afterword as someone whose professional career is not necessarily at stake in making substantive claims about biology, I hope that the reader will allow me

*Director, Laudato Si' Research Institute, Campion Hall, Oxford University, UK.

to indulge in speculation about where it might go next in a way that may be less permissible for those who are embedded in institutional structures that might not allow such claims to be made. At the same time, I would like to reassure the reader that these claims are no purple passages of the type sometimes found in popular science labelled as scientism, but a willingness to respect the integrity of biological ways of thinking while wanting to probe questions about potentially new directions, ones where the non-specialist, at least, also has a rightful say in its unfolding.

To claim that biology is changing does not mean, of course, that the overall fashions in biology are totally new. I was particularly struck by Derek Gatherer's comments on how the research focus on molecular biology in the 1980s transformed eventually into systems biology, but of course holism of some sort has been part of biological history for centuries. The dynamism of change within biology is more like a moving spiral, returning back to some historically situated ideas, but edging forward bit by bit. Like Gatherer, I was acutely aware of the social and funding pressure towards molecular biology in the Thatcher era, and at the time, ecologists were often seen as marginal to cutting edge biology. I too, heard the same joke as an undergraduate in the natural sciences at Cambridge about biochemists blending everything in a bucket and hoping something more or less like what happens in the real life will show up. As a plant physiologist I felt caught in the middle, unable to commit to molecular biology, not least because of that sense that parts could not explain everything, and yet at the same time dissatisfied with the generally descriptive approach to ecology.

The new wave in biotechnology that took over whole departments, funding botany departments from industry and often in accordance with specific practical needs rather than the quest for knowledge, also put those who went into biology for its own sake and out of sheer curiosity about how things work on a back foot. Now, I am not saying that it was impossible to bring creative knowledge to bear in highly applied areas, or that working on such problems did not have research projects that sometimes led to important new theoretical insights, but that the pressure to be a commercial success somehow overlaid other epistemological considerations and undercut biology's freedom to develop. It was, in other words, a serious restraint. Today, few could perhaps conceive of biological research *without* some private backing. Nonetheless, I was mildly surprised that this did not seem to come up at all in this volume, and the only explanation I can think of is that there has been an accommodation and adaptation to working with both private and public funding. Further, perhaps the fears that some of us had some thirty years ago that biology as we knew it would cease has not been realised, as Pope Francis has pointed out, scientific creativity cannot be suppressed,[a] even where funding is directed towards practical aims. In addition, the private funders may also be more flexible now than they once were about funding research that does not necessarily lead to immediate results.

The example of work by Ottoline Leyser, Director of the Sainsbury Laboratory in Cambridge, is a case in point. Leyser is clearly driven by curiosity about how *Arabidopsis* develops through time. The conceptual language of meristem, growth, perturbation etc. brought back good memories to me of the time when such discussions were the bread and butter of my own phase as a practising biological scientist. And, her discussion of levels of analysis and the importance of clarity in thinking through where the slice of research under investigation was situated in the grander scheme of things is critically important, not just for developmental systems biology, but more broadly for the task of any biologist. Such appreciation of the complexity of the system and the extent to which our own knowledge is limited by the specific problem being investigated constantly challenges a scientist to stay humble before the face of a great deal of ignorance. Admitting to such difficulties is not always popular in public renditions of science,

---

[a] Pope Francis insists that science needs to be in service of the common good, and, in so far as science has applied elements, I believe this is entirely correct. So, 'If an artist cannot be stopped from using his or her creativity, neither should those who possess particular gifts for the advancement of science and technology be prevented from using their God-given talents for the service of others'[1] (§131).

perhaps because it might look like a sign of weakness. However, it seems to me that it is important not to give a false impression that more progress is being made than is actually the case. The gaps in knowledge and the rationale for trying to fill those gaps is what makes science the intellectually challenging and satisfying, yet also at times frustrating activity that it is.

Another important development that is reflected in many of the chapters represented in this book is the importance of different disciplines within biology talking to each other. Developmental biology is a good example. In evolutionary biology Developmental Systems Theory is one end of the spectrum while the Modern Synthesis is at the other end. There are heated debates among evolutionary biologists as to how 'new' such theories are relative to the neo-Darwinian paradigm. I am sympathetic with David Depew's push towards a developmental approach to evolutionary thinking, and his resistance to the resolute materialism of the likes of Richard Dawkins, Stephen Pinker and Daniel Dennett, along with other more specific problems associated with the Modern Synthesis. I can also see how such an approach is aligned in many respects with at least a form of process philosophy. However, the thought of A. N. Whitehead is not necessarily in view here, not least because he understood processes to have their origins in the physical structures of the world that may not be sufficiently satisfying for biologists, unless, of course, the non-life/life boundary issues also come into view. I think, however, that Depew is making a significant claim in saying that in evolutionary biology we are witnessing an unfolding paradigm shift, one that has tremendous public significance. Part of the difficulty is that those who are still stuck, as he suggests, in 19th century frameworks still have the loudest public voice. Reclaiming a different public voice for an alternative way of understanding the task of evolutionary biology has practical implications, for the newer theories such as the extended evolutionary synthesis approach is, as I have argued in a number of different places, rather more compatible with theological thinking.[2] Similar attempts to challenge reductionism and affirm a paradigm shift in evolutionary thinking are laid out by Ilya Gadjev in his chapter that deals with more specific issues of genetics in relation to epigenetics and significance of such trends to how we think about what it means to be human.

The engagement of developmental ideas with evolutionary ones illustrates another important point: namely, that growth points in biology are often at the intersection of two sub-disciplines, rather than at the core of the discipline itself. Ecology, for example, like developmental biology, is now finding a robust role in relation to evolutionary biology, as Richard Gunton and Francis Gilbert discuss in their chapter. The extended evolutionary synthesis concepts championed by scholars such as Jeremy Kendal,[3] or Kevin Layland,[4] show the importance of ecology but without losing all sight of the contribution of natural selection. What is still left of the Modern Synthesis is a humbled and pruned variety of that hypothesis. I am sympathetic, too, to the idea that ecology should be given its proper place as no longer the Cinderella subject (though the authors do not use that explicit terminology), relegated to the status of a soft science or even afterthought in a way that has, at least in the past, been habitual in biological education programmes. At the same time, it is worth pointing out the value of what I would call a mutual interlacing of two or more areas of knowledge, so that just as evolution becomes more expansive and self-critical when it takes up ecological systems as part of its brief, so too, ecology enlarges its own understanding when evolutionary history becomes integral to its way of approaching the study of living communities.

One of the next large challenges for evolutionary biology from my own work with evolutionary anthropologists and archaeologists is to find better ways to relate biological and cultural evolution, also touched upon in Gadjev's chapter. And, as Elizabeth Jones and Michael Ruse are aware, when it comes to speaking about the evolution of humanity there really is no such thing as a 'private science of human nature'[5] (p. 108). Rather than speaking of the historical and public aspects of these issues, however, that they lay out clearly in their chapter, I am most concerned here with current debates *within* evolutionary anthropology, not least because I believe they also have significance for how to think about religion. My premise is the following: if it can be shown that anthropologists are becoming more open to consideration of what it means to be religious, or perhaps in evolutionary terms, rather better, what it means to be spiritual, then, in so far as they are also incorporating in an important way biological understanding, so this opens a crack in the biological door that resists the possibility that theology may actually have something important to say to science. This takes the debate beyond simple consonance to something rather more adventurous. For pure consonance is unlikely, just as it is unlikely across scientific disciplines or even between them, as the ongoing dynamism within biology shows. I am not suggesting that all readers of this book will necessarily agree with this claim. However, what I am arguing for is that keeping an open mind as a scientist, through the lens of anthropology, at least makes such a possibility a relevant one.

Cultural evolution is distinct from straight biological evolution in important ways, especially if the biological claims are understood in terms of pure reproductive success. Here we arrive at a border in anthropology that is also relevant for thinking through the boundaries between biological sciences and other areas of knowledge, namely, the complex relationships between the quantitative and qualitative, though what is really interesting is that cultural evolution also displays a combination of both methods. Social anthropologists have often been suspicious of quantitative methods, as they seem to miss out the rich and dense textured lives of real people living in specific communities. Biological anthropologists, on the other hand, have understood qualitative approaches as missing out on empirical or evidence-based research, meaning that conclusions reached by more qualitative methods are limited, at least as far as they can be termed scientific. In practice, of course, the situation is rather more complicated than this division implies. The design of a good qualitative research project also has quantitative research elements as part of its background. For example, there is a clear and often statistically based rationale in choosing this particular area of study and that particular community. Further, any observations made that are more qualitative can be appraised through quantitative methods, or a mixed method approach, as John Dunce from the Max Planck Institute for Evolutionary Anthropology has helpfully laid out in his study of two different indigenous communities.[6] At the same time, Matei Candea's fascinating research on animal scientists studying the behaviour of Meerkats in the Kalahari Meerkat project[7] is worth a comment. He found that the scientists' relationships with those animals are mixed: as scientists they are objective, but that does not mean that from a day to day relational perspective they do not perceive those animals as subjects.

In his Layton Dialogue lecture,[8] Candea spelt out how both quantitative and qualitative methodologies are in fractal relationship with each other, behind quantitative data sets are qualitative ideas and behind qualitative ideas is the search for more quantitative assurance. The fashion for big data in biology should also not obscure the difficulty in arriving at clear conclusions simply from correlational data sets. Anthropology helps other sciences, including the biological sciences, understand the very human practice that makes science what it is, but without taking away from the importance of the search for secure forms of knowledge.

What is also really interesting, of course, is that public understanding has not really caught up with the trends towards holism in biology, as exemplified in the way reductionism about the human person, be it genes, brains or intelligence tends to seem more attractive compared with more complex alternatives. Harris Wiseman illustrates this point beautifully in his twin chapters on the brain where neurocentrism seems to be the latest fashion in public reception of science. There are other important examples to consider in this context too, such as the importance of identity in the case of dementia and the end of life. The question of how we understand human identity towards the end of life taken up in Will Beharrell's chapter is going to impact on most of those reading this book, given the current life expectancy in the Western world. The question of the impact of the environment on life systems, including the microbiome touched on in Wiseman's chapter leads onto another extremely important area that is likely to become a growth area in biology, namely, that of mutualism and symbiosis. This is also a contributor to the shift towards a more holistic way of perceiving biology, but it has profound implications for how to think about evolutionary biology as well.

For example, in a fascinating book on computer modelling of evolution, Richard Watson argues that certain types of complex system, which could not evolve from a model using standard gradualist frameworks, are possible through what he terms composition, that is, bringing together two separate genetic lineages.[9] One example of this is already part of evolutionary theory and familiar to most people — sex. The other is symbiotic encapsulation.[10] He proposes that standard evolutionary theory by natural selection amounts to a hill-climbing model of optimisation, which refers to relative fitness. However, that does not mean that it is the only form of optimisation. Critically he spells this issue:

> The real distinction between compositional evolution and gradual evolution is not that one is nongradual and one is gradual, but one is compositional and the other is linear — that is what I refer to as the 'gradualist framework of evolution' assumes the linear sequential accumulation of undirected genetic changes (whatever size they may be).[9] (p. 274)

The actual compositional events themselves may or may not be beneficial associations. The point is that there are *some conditions* where, even allowing for non-bias in compositional events, on average such compositional processes are *more likely to be beneficial* when compared with undirected variation of both beneficial and nonbeneficial types produced by mutations. Each compositional event may have prior compositional events preceding it, so the computational models 'support the possibility that symbiosis potentially produces a fundamentally different source of evolutionary innovation from the accepted norms of evolutionary change'[9] (p. 276). The overall result is that what might appear to be impossible complexity in an evolutionary sense becomes feasible through a compositional process. Watson is correct, in my view, to have running in parallel *both* evolution by natural selection as well as this compositional approach, rather than necessarily replacing one with the other. However, what it does show up in a fascinating way is that evolutionary theory is yet more complicated still, and the standard approach to evolution by natural selection is insufficient.

But what might any of this have to do with a religious sense? Agustin Fuentes gave a wonderful series of Gifford lectures in 2018 that I have had the privilege to read prior to publication.[11] In these lectures he argues that trying to understand why we have come to hold beliefs the way we do is a fundamental aspect of what makes us distinctly human. He also notes that there are 5.8 billion people around the world who identify as religiously affiliated. However, all religions and religious institutions are very recent in evolutionary terms. But that does not mean that an experience of the transcendent is confined to such a time frame, or that hints in the evolutionary record cannot point to the possibility of experience of the transcendent. In some fascinating work with macaques, Fuentes has shown that on some, but repeated, occasions macaques will be arrested by a scene that can only be described as stunningly beautiful, and there seems to be no other explanation for why they stood still in their tracks. Similar examples of chimpanzees watching waterfalls have been noted by Jane Goodall.[12] There needs to be some care, of course, in interpreting such observations. What it does show is that the possibility of a deep phylogenetic origin to the capacity to experience wonder is likely. What it does not show is whether these experiences are necessarily exactly the same as the kind of wonder filled experiences humans feel when confronted with beautiful imagery. And wonder, of course, is also distinct from religious awe, even if the two are related.[13]

The puzzle of why religion emerged at all is a difficult one for biologists to come to terms with on a purely evolutionary level. Rather than dismiss religion as a 'spandrel' arising out of a complex brain, an alternative, rather more positive explanation, at least in evolutionary terms, is that religion helps to solve the problem of how to deal with free riders in densely cooperative societies.[14] However, there is something dissatisfying about this theory as a full explanation, especially when it comes to reflection on hunter–gatherer groups. There is no evidence, for example, that there is a need to control free riding through more authoritative means using moralising gods in small-scale societies. As Christopher Boehm has shown, the shame associated with 'cheating' on other group members is sufficient to curtail anti-social behaviour.[15]

An alternative, and one that I am currently exploring with evolutionary archaeologist Penny Spikins, is that as the exchange of goods gradually took place over longer and longer distances, the possibility of an absent other eventually made the possibility of an absent transcendent being a believable one. As a theologian I would argue that these experiences of the transcendent were real, rather than reflecting a convenient or even evolutionary significant illusion. Spikins concurs that there is no way that the scientific evidence can distinguish the truth claims or not of belief in gods. But the fact that there is interest in and affirmation of the importance of the transcendent in the lives of humans is, I suggest, at least a step in the right direction towards healing the gulf that still exists, especially in the public mind, between science and religious belief. This book has hinted at such a dialogical approach in some places. It is my hope that this aspect will be built upon and elaborated in research in this century, not least for the sake of the public understanding of science. For if, as anthropologists have shown, 83% of the world's population are religious, then is it really surprising that science is not taken seriously enough if such considerations are routinely bracketed out from any discourse?

Finally, in my closing remarks, I'd like to stress how important it is that biologists also take some responsibility for the ethical aspects of the work that they do. This is hinted at in the chapter by Michael Reiss on food where a discussion of animal welfare crept into his discussion. Rearing animals in confined conditions in highly artificial environment and forced to suffer for the sake of producing cheap meat is something that biologists need to be just as concerned about as members of the public.

Ethologists, for example, like Marc Bekoff, have found that social animals are extremely sophisticated in their neurochemistry, social emotions and intelligence.[16] Indeed, encountering his work made me even more aware of the importance of social emotions in other animals and their importance for human evolution as well. If their lives are so interlaced with our own, what right do we have to treat them as objects that can be treated as if they are automata?[2] Such insights have not yet reached the general public, given the habitual practice of buying meat where the animals concerned are known to have endured all kinds of harm. This is not just a matter of vegetarianism becoming an impossible ideal, but of being responsible for the products that we buy, knowing the implications of the scientific knowledge about the natural world. New gene technologies can also, of course, manipulate animals so that they are less likely to undergo stress in confined conditions.[17] I consider such manipulations a misuse of science, rather than its responsible use. We need, as Steven Yearley suggests in his chapter, citizen science.

References

1. Pope Francis. (2015) *Laudato Si': On Care for Our Common Home.* Catholic Truth Society, London.
2. Deane-Drummond C. (2014) *The Wisdom of the Liminal: Evolution and Other Animals in Human Becoming.* Eerdmans, Grand Rapids.
3. Kendal J, Tehrani JJ and Odling-Smee FJ. (2011) Human niche construction in interdisciplinary focus. *Philos T Roy Soc B* **366**(2011): 785–792.
4. Laland KN, Odling-Smee FJ and Feldman MW. (2000) Cultural niche construction and human evolution. *Behav Brain Sci* **23**(2000): 131–175.
5. Jones E and Ruse M. (2019) Human evolution: From fossils to molecules, reductionism to holism. *(This volume.)*
6. Dunce J. (2018) Two paths to cross-cultural competence and one to cultural sustainability. *In preparation*, discussed at Layton Dialogue with Matei Candea: Is quantity just a matter of quality? Department of Anthropology, 12th December 2018, Durham University, UK.
7. Candea M. (2018) The two faces of character: Moral tales of animal behaviour. *Social Anthropology* **26**(3): 361–375.

8. Candea M. (2018) Is quantity just a matter of quality? Layton Dialogue with John Dunce, Department of Anthropology, 12th December 2018, Durham University, UK.

9. Watson RA. (2006) *Compositional Evolution. The Impact of Sex, Symbiosis and Modularity on the Gradualist Framework of Evolution.* MIT Press, Cambridge, Mass.

10. Deane-Drummond C (*in preparation*) Symbiotic wisdom: Recovering a memory of deep history. *Theology and Science*, Special issue on Mutualism, edited by Davison A.

11. Fuentes A. (2018) *Why We Believe: Evolution, Making Meaning and the Development of Human Natures,* Gifford Lectures, 26th February–8th March, University of Edinburgh.

12. Deane-Drummond C. (2015) Wonder and the religious sense in chimpanzees. In: Petersen D and Bekoff M (eds.) *The Jane Effect: Celebrating Jane Goodall,* pp. 225–227. Trinity University Press, San Antonio.

13. Deane-Drummond C. (2004) *Wonder and Wisdom: Conversations in Science, Spirituality and Theology.* DLT, London.

14. Johnson D. (2014) *God is Watching You.* Oxford University Press, Oxford.

15. Boehm C. (2012) *Moral Origins: The Evolution of Virtue, Altruism and Shame.* Basic Books, New York.

16. Bekoff M. (2014) *Re-Wilding Our Hearts: Building Pathways of Compassion and Co-Existence.* New World Library, New York.

17. Deane-Drummond C. (1997) *Theology and Biotechnology: Implications for a New Science.* Geoffrey Chapman Cassell, London.

Index